The GRAFT HYBRID

UNIVERSITY OF PITTSBURGH PRESS

The GRAFT HYBRID

CHALLENGING TWENTIETH-CENTURY GENETICS

Matthew Holmes

Published by the University of Pittsburgh Press, Pittsburgh, Pa., 15260
Copyright © 2024, University of Pittsburgh Press
All rights reserved
Manufactured in the United States of America
Printed on acid-free paper
10 9 8 7 6 5 4 3 2 1

Cataloging-in-Publication data is available from the Library of Congress

ISBN 13: 978-0-8229-4793-6
ISBN 10: 0-8229-4793-5

COVER ART: Florentine Bizzaria in Antoine Risso and Pierre-Antoine Poiteau, *Histoire naturelle des orangers* (Paris: Mme Herissant Le Doux, 1818), plate facing p. 107.

COVER DESIGN: Alex Wolfe

For FLORENCE *and* HAROLD HAGUE

CONTENTS

ACKNOWLEDGMENTS

WHEN I LOOK BACK ON THIS MULTIYEAR PROJECT, I FEEL SOME ALARM that I cannot accurately recall all the friends and scholars who helped me through its difficult stages. Consequently, these acknowledgments are likely to be incomplete and I apologize to anyone I have missed.

The idea to write a book on graft hybrids first emerged during my Arts and Humanities Research Council–funded research at the School of Philosophy, Religion and History of Science at the University of Leeds. I would never have come across graft hybrids, or felt able to adequately address them, without the support of my wonderful supervisors, Gregory Radick and Tina Barsby. As both this book and its author have traveled to new places, Greg has been a constant source of stability to whom I owe a great deal of thanks. My sincere gratitude also goes to Graeme Gooday and Jon Agar. The History and Philosophy of Biology reading group at Leeds has long provided a friendly and intellectually stimulating space and was an invaluable place for me to develop and test the manuscript in its early stages, for which I am extremely grateful.

The bulk of this book was written during my position with Inanna Hamati-Ataya on her European Research Council–funded project ARTEFACT, at the Centre for Research in the Arts, Social Sciences and Humanities at the University of Cambridge. I am grateful to Inanna for affording me the time and space to make this book a reality.

My thanks to the archival staff at the British Library, Cambridge University Library Special Collections, Edinburgh University Library Special Collections, the John Innes Centre, the Library of the American Philosophical Society, the Linnean Society Archives, the Needham Research Institute, and the Science Museum Library and Archives (London), without whom this book would not have been possible.

Several people have generously supported this book project in various ways over the years. I am especially grateful to Helen Anne Curry, Dolly Jørgensen, David Munns, Jan A. Witkowski, and all my friends and colleagues in Leeds, Cambridge, and Stavanger.

ACKNOWLEDGMENTS

Many thanks to the two anonymous reviewers for their invaluable feedback, which greatly improved the book. Their interventions made the illegible somewhat more legible and removed some bad analogies and over-dramatization. My thanks to Abby Collier at the University of Pittsburgh Press, with whom it has been a pleasure to work.

Finally, to Annie—I would write something about us being like two grafted flowers, but that would be another bad analogy. Instead, I will just say thank you for everything.

The GRAFT HYBRID

FIGURE 1. Florentine Bizzaria. The Bizzaria, which contained elements of citron and orange in a single tree, was first reported in seventeenth-century Florence, Italy. Scientific interest in the plant peaked in the early twentieth century, as its existence suggested that plant hybridization could occur through grafting. Antoine Risso and Pierre-Antoine Poiteau, *Histoire naturelle des orangers* (Paris: Mme Herissant Le Doux, 1818), plate facing p. 107.

INTRODUCTION

The Florentine Bizzaria

—✑

IN 1674, A PLANT THAT SHOULD NEVER HAVE EXISTED APPEARED ON THE outskirts of Florence. The aptly named "Florentine Bizzaria" was a hybrid being (fig. 1). It had been made by grafting—the physical joining of one organism's tissue to that of another—a Florentine citron and a sour orange.[1] Accounts of why the Florentine gardener Pietro Nati had chosen to cross the boundaries between species are varied and contradictory, although at least one held that his foray into transspecies engineering was a mistake.[2] The plant itself certainly looked poorly designed. If you cut into its fruits and dared to taste them, you would find that some segments were bitter, while others tasted of orange. Although it sounds like a remarkable plant, the arrival of the Bizzaria in seventeenth-century Florence probably had little impact upon the city's populace. The idea that new plants could be created through grafting was commonplace in early modern Europe, although questions remained over their true nature.[3] Many accounts of the Florentine Bizzaria instead date to the early twentieth century, when the claim that hybrid organisms could be made by grafting attracted intense scrutiny. As one participant in the scientific debates over these "graft hybrids" would explain, the revived interest of his contemporaries in artificially conjoined plants and animals was "no doubt in association with the revival in that of genetics brought about by the 'rediscovery' of Mendel's work."[4]

3

Plant grafting is generally performed by attaching a branch (scion) of one plant to the stem (stock) of another. In horticulture, its primary purpose is to propagate plants. By attaching several scions of one plant to the stock of another, you are essentially "cloning" the former.[5] This operation can also combine useful characteristics from different plant varieties. When the French wine industry faced a devastating insect pest in the mid-nineteenth century, botanists found that grafting French grapevines onto resistant rootstock from North America, where the insects had originated, could save the crop. This solution was not without some controversy, as French vineyard owners were concerned that the "inferior" flavor of American wine might seep across the barrier between the plants (a site known as the "graft junction") and pollute their grapes. Eventually, this grafting became the norm and was even celebrated.[6] As the twentieth century dawned, however, grafting came under increased scrutiny from biologists. They had almost the same concern that the French viticulturists had, in that they wished to know whether grafted plants could exchange characters—ranging from disease resistance to fruit color—across the graft junction. Nor were they limited to grafting only plants. Surgical advances meant that organs could be transplanted between farm or laboratory animals. If characters could be swapped between organisms to create hybrids, which in turn could pass these mixed characters on to their offspring, it would transform how twentieth-century biologists understood heredity and open new economic possibilities for agriculture.

This book is about the graft hybrid and its turbulent relationship with genetics in the twentieth century and beyond. Graft hybrids offered many things: the possibility of fantastical new species, economically important new crops, and insight into alternate and unknown forms of heredity. Yet they repeatedly clashed with certain interpretations of genetics. By reconstructing their history, in this book I make three claims about twentieth-century heredity and biological experimentation. The first is that the very existence of graft hybrids challenged emerging techniques and concepts in genetics, shaping the discipline across the twentieth century. In this period, genetics was still a nascent science. Its grasp on biology and agriculture was not assured. Graft hybrids—living, breathing plants and animals beyond contemporary understanding—presented a real conundrum, or a real threat, to genetics. The years after the "rediscovery" of Mendelian genetics in 1900 have been described as a period "when hybridizers sought to fathom the limits and implications of Mendel's generalizations."[7] Graft hybridizers claimed not only to have discovered limits to Mendel's laws but also that graft hybrid-ization could break these limits. After all, there were technically no obvious boundaries as to what organism could be grafted onto another.

The second argument I make in this book is that the twentieth-century decline of graft hybrids was gradual and overrated. Most scholarly attention to graft hybrids has been focused on two periods: the nineteenth century, a time when Charles Darwin found himself focused on plant grafting to un-derstand heredity and development; and the appearance of the graft hybrid in the Lysenkoist biology of the Soviet Union from the 1930s onward. The intervening decades of the twentieth century, however, are equally fasci-nating. During this time the scientific discipline of heredity was gradually emerging, with no concrete consensus over its objectives or limitations. Fur-thermore, heredity was emerging against a "background of the formation of nation-states, with their centralized bureaucracies, capitalist economies, and imperialist aspirations."[8] Into this heady mix was thrown the graft hy-brid. An intense debate over its true nature occurred just prior to the First World War. During the interwar years, several prominent supporters of graft hybridization could be found among European botanists and zoologists. The existence of graft hybrids was disputed, but by no means debunked. One important theme of this era was the failure of repeated attempts to bring experimentally produced graft hybrids into the scientific mainstream. These failures were often contingent or happenstance, with a series of lost specimens contributing to their mystery.

Third, and finally, I argue that graft hybrids were the most compel-ling part of Trofim Lysenko's biology in the Soviet Union. Defenders of Lysenkoism in the West, including the British biologists J. B. S. Haldane and Anne McLaren, pointed to graft hybrids when challenged to provide proof of Lysenko's doctrine. Unlike many aspects of Soviet biology, there was a genuine sense of the unknown surrounding graft hybrids. As I will show in this book, their existence was debated by scientists for much of the twentieth century, with no clear answers emerging. Thanks to this tradition, Lysenkoists' claims to have created graft hybrids were not unbelievable. In contrast to many of Lysenko's "experiments," graft hybridization was at-tempted by researchers across the Cold War world. Detailed accounts of graft hybrid plants and animals arose from Eastern Europe and China. Some of these experiments could not be replicated, while others left unexplainable results. Unfortunately for graft hybridizers, however, graft hybrids came to be rejected in genetics partly as a result of their association with Lysenkoism.

As Lysenko's attacks on genetics in the Soviet Union mounted, Western geneticists doubled down on the importance of classical genetics, dismissing Lysenkoism as pseudoscience. By the mid-twentieth century, any association with graft hybridization was to be avoided.

Graft hybrids were some of the strangest and most controversial beings that might never have existed. But before we can delve into their equally strange history, we must first explore the historical context in which these organisms arose to challenge narrow conceptions of heredity. The modern conception of graft hybrids developed in the late nineteenth century amid a sea of other hereditarian concepts and theories, including the biology of August Weismann and the rediscovery of Mendelian genetics at the dawn of the twentieth century. To understand some of the controversy surrounding graft hybridization, we must also investigate our current understanding of its role in the "Lysenko affair." In short, the twentieth-century history of graft hybrids touches upon some of the most important episodes in the history of biology.

The Graft Hybrid

Charles Darwin coined the term *graft hybrid* in his 1868 book, *The Variation of Animals and Plants under Domestication*, to describe an organism created through grafting that was identical to a typical sexual hybrid. It is possible, he wrote, that "two distinct species can unite by their cellular tissue.... Such plants, if really thus formed, might be called graft hybrids."[9] Darwin, however, had not come up with a new concept. He had simply named and defined an old idea in contemporary scientific language. Grafting was familiar to the ancient Greeks, although Aristotle struggled to fit grafted plants into his philosophical system, as they represented an awkward intersection between the artificial and the natural.[10] In the 1930s the classicist and amateur botanist Arthur Stanley Pease suggested that ancient writers, including the European "father of botany," Theophrastus, were wary of grafting between different species of plant lest they breach the "principle of limitation" that divided species. Pease, however, was reading the anxieties of his own time surrounding graft hybridization into classical texts.[11] Roman botanical texts and poetry had freely engaged with the "fantasy of unlimited transplantation."[12] By the nineteenth century, graft hybrids and their role in heredity became the subject of systematic inquiry by horticulturalists and botanists. Classical texts still made the odd appearance. In 1842 the editors of the *Gardeners' Chronicle*, a popular British horticultural magazine, were alerted to an outlandish claim in a rival publication. Someone had claimed to have produced a rare and valuable yellow rose by grafting a common red rose onto a broom (a type of flowering shrub). The *Gardeners' Chronicle* did not stand for such nonsense, suggesting that believers of such stories must have been reading too much Virgil or Columella, a sad indictment of the "nature of an English education."[13]

Early nineteenth-century commentators on what would soon be termed "graft hybrids" possessed a certain skepticism. Various accounts of such hybrid plants were gathered by Antoine Risso, former professor of physical and natural sciences at the Lycée de Nice, in partnership with Pierre-Antoine Poiteau, chief gardener of the Royal Nurseries of Versailles. Together they produced the monumental *Histoire naturelle des orangers*, published in 1818. In it, they refuted most claims that grafting could create new species. The Maltese orange tree, for example, was popularly claimed to be the product of a scion from an orange tree grafted onto a pomegranate shrub. Risso and Poiteau dismissed this account as an absurdity that had been disproved through experiment.[14] Yet one botanical puzzle eluded them. "Here is the most singular and curious tree of all the vegetable kingdom," they wrote of the Florentine Bizzaria. They accepted that the tree had likely been created by grafting but had no way of explaining its strange appearance. Other unexplainable botanical oddities also caught their attention. The duo located an orange tree in Nice that bore both bigarades and oranges and brought forth a mixture of red and white flowers.[15]

By 1868, Darwin was able to give two examples of famous graft hybrids that would have been familiar to his readers. One of them was the Florentine Bizzaria; the second was a tree known as the *Cytisus adami*, which Darwin described as a "form of hybrid intermediate between two very distinct species"; namely, the common and the purple laburnum.[16] The *Cytisus adami* had been created by Jean-Louis Adam, a Parisian nurseryman, before being brought to the attention of European botanists in 1830. One of the first on the scene was none other than Poiteau, who published several descriptions of the plant in the journal of the Parisian Society of Horticulture.[17] News of the unusual plant spread rapidly, with it winning several supporters. One British advocate was the clergyman and naturalist William Herbert, who claimed in 1840 that the existence of the *Cytisus adami* raised new and exciting possibilities. "If I am right in my notion," declared Herbert of his support, "it opens a field for the horticulturalist to produce hybrid plants which perhaps could not be obtained by seed."[18] Herbert's support demonstrates the appeal of graft hybridization to the practical plant breeder. If real, such organisms could bypass the natural limits faced by sexual crosses. In an ambition reminiscent of modern genetic engineering, almost any species could be combined with any other.

Darwin had begun his own investigation into the *Cytisus adami* in 1847, convinced that the tree could provide living evidence to support his belief "that the entire body of the organism had a role to play in determining heredity."[19] Although Darwin was not able to re-create the *Cytisus adami*, he nonetheless used *The Variation of Animals and Plants under Domestication* as a vehicle to introduce his own theory of heredity—pangenesis—to his Victorian peers. Put simply, pangenesis is the idea that each organ or cell of the body throws off a minute copy of itself. These copies, or "gemmules," then congregate in the sexual organs and are the means by which the physical characteristics of parents are passed on to their offspring.[20] If the units of heredity reside within the cells of living bodies, a graft hybrid would offer powerful evidence in favor of pangenesis. By taking the body part of one organism and surgically grafting it onto the body of another, the appearance of any characteristics resembling the grafted part in subsequent offspring would indicate that the wider body, not just the sex cells, could influence inheritance. Darwin was also able to present several accounts from gardeners and horticulturalists claiming that hybrid apples and roses could be made using grafting. Though compelling, however, these anecdotal stories were not enough. Darwin admitted that "it is at present impossible to arrive at any certain conclusion with respect to the origin of these remarkable trees."[21]

Shortly after the publication of *The Variation of Animals and Plants under Domestication* in 1868, Darwin received a letter from Friedrich Hermann Gustav Hildebrand, a German professor of botany at the University of Freiburg. Hildebrand had read Darwin's book and believed that graft hybridization could explain some strange results from his own experiments with potatoes. Hildebrand had grafted sprouts from white potatoes onto red potatoes, from which he managed to grow two bushes. Some of the potatoes brought forth by these bushes, he informed Darwin, "held the middle between the red and white potatoes: they were red and scaly at the one end, white and smooth at the other and in the middle smooth and white with red stripes."[22] This experiment convinced Hildebrand that graft hybrids did exist. However, his efforts to replicate his results had been unsuccessful. In another letter to Darwin, Hildebrand apologized for this failure. His two bushes had stopped producing potatoes and other attempts at grafting went nowhere.[23] Darwin directed his own gardener to graft differently colored potatoes together but did not come up with anything representing a hybrid.[24] Darwin and Hildebrand had run into a problem that would afflict research on graft hybrids for the next century: graft hybrids could not be reliably produced or replicated.

In the last quarter of the nineteenth century, an elderly and increasingly frail Darwin enlisted the aid of a young and enthusiastic naturalist to assist his search for a graft hybrid. George Romanes, like many other Victorian gentlemen of science, was engrossed by pangenesis. Romanes had been busy testing the theory by removing the ears of rabbits and other mammals for surgical grafting. Darwin encouraged him to abandon this approach and instead conduct grafting experiments on plants, particularly potatoes. From 1875 to 1880, Romanes grafted numerous species of plant together: potatoes, beets, onions, dahlias, peonies, and carrots. Regrettably, success was not forthcoming. Plants were lost to disease; grafted plants decayed or separated from their hosts. All the resulting seeds displayed the characteristics of only a single parent.[25] Results from other thinkers in the life sciences also spelled bad news for Darwin's theory of pangenesis. In 1871 the English polymath and eugenicist Francis Galton had found that transfusing blood from one variety of rabbit to another resulted in no "alteration of breed" in their offspring, demonstrating that "the doctrine of Pangenesis, pure and simple, as I have interpreted it, is incorrect."[26]

Galton was not the only skeptic. Charles McIntosh, a Scottish horticulturalist who had worked in the gardens of European monarchs and the British aristocracy, took a thoroughly practical approach to the question of graft hybrids. Drawing upon his extensive experience as a well-traveled gardener, McIntosh related several botanical observations that would cause the inquisitive mind to doubt tales of hybrids created by grafting. One could be seen by simply cutting into the point where two grafted plants were joined, at what was later referred to as the *graft junction*. McIntosh described how the two plants maintained their own distinctive "layers," which could be easily peeled apart from each other by hand. This indicated that grafted plants were mechanically pressed together, with no evidence of more fundamental intermixing or blending, an argument that would be formalized in the twentieth century via the formation of the chimera hypothesis. He also noted that many hundreds of trees were propagated through grafting at any one time. The vast majority of these plants maintained their distinctive identity and produced their own fruit.[27] If some kind of hybridization happened at all, it was rare.

These nineteenth-century interactions with graft hybrids display characteristics that would emerge time and again over the course of the twentieth century. Graft hybrids—and their existence or nonexistence—were bound up with fundamental questions about how heredity worked. In pursuit of answers to this mystery, little or no distinction was made between plant and animal graft hybrids. There were hints of the power that graft hybrids, in the form of new and previously unimaginable organisms, could grant their creators. Grafting could potentially burst through the species barrier, providing new plants and animals for agriculture that could not be obtained through other means. As for Darwin, he worked until the end of his life in an ultimately futile attempt to acquire a graft hybrid. If he had succeeded, then twentieth-century debates over the nature of heredity might well have been conducted on rather different grounds.

The Weismann Barrier

By arguing that hereditary material was produced throughout the body, Darwin's theory of pangenesis incorporated the age-old belief in the inheritance of acquired characters. The theory of acquired characters is most commonly associated with the French zoologist Jean-Baptiste Lamarck, who, in the early nineteenth century, proposed that parts of a living being could be gradually modified in response to changing environmental conditions, and that these modifications, which occurred during the lifetime of a single organism, could then be inherited by its offspring. His subsequent celebrity led to the emergence of "Lamarckians," a general epithet for those who stressed "the evolutionary role of individual variations that emerged during the life of an organism in response to environmental stress."[28] The most famous example given by Lamarck was the neck of the giraffe. By continually stretching its body to reach the scarce leaves on trees, went the theory, a giraffe had lengthened its neck and forelegs. These changes were inherited by its offspring, who repeated the process again and again across the generations until the giraffe had achieved its height.[29] For now, we can leave aside the question of what biological mechanism could explain this change and its inheritance. The key takeaway of this theory is that the body could adapt to changes in its environment, and that these changes could potentially be passed down on to its offspring.[30]

In the mid-nineteenth century, the inheritance of acquired characters was a relatively common idea, its presence arousing none of the ire directed at Lamarckism by modern biology. But scientific inquiries into the nature of the cell were already beginning to separate heredity from the development of the body during an individual's lifetime, thus undermining possible mechanisms by which alterations to the somatic (body) cells induced by environmental change could be inherited. In 1858 the German physician Rudolf Virchow theorized that there was "a division of labor" between the cell nucleus, which contains the chromosomes, and the cytoplasm, the liquid interior of the cell in which its organelles sit.[31] An elegant experimental demonstration of the importance of the nucleus in heredity was conducted by German zoologist Theodor Boveri in 1889. While based at the Zoological Station in Naples, Boveri removed the nuclei from the eggs of a genus of sea urchin. He then fertilized these eggs with sperm cells from another genus, which still contained their nuclei. The subsequent sea urchin larvae resembled only the latter. This result proved that it was the nuclei, not the cytoplasm, that shaped the developing larvae. The heredity process seemed to be confined to the nucleus of the cell. Boveri's experiment was well received, with the geneticist Thomas Hunt Morgan translating his report into English.[32] Morgan would later argue against Lamarckism and suggest that the chromosome was the site of the gene.

At the same time that Boveri was asserting the primacy of the nucleus, his countryman August Weismann was coming to a similar realization. Trained in medicine before turning to zoology, Weismann had once idolized Lamarck and Darwin. Yet in 1883 Weismann asserted that embryological "overgrowth" was a special phenomenon unique to the germ (or sex) cells. In 1885 he argued for the continuity of this "germ-plasm," stating that hereditary characters were passed down from parent to offspring through the germ cells alone. The somatic cells that make up the rest of the body had no role in this process. Since the germ cells of animals were distinct and isolated from changes in the somatic cells, this left no room for a mechanism by which the inheritance of acquired characters could occur. Weismann methodically rejected other heredity theories of his time, including Darwin's pangenesis, which suggested that acquired characters were transmitted from the body to the germ cells. Weismann's rebuff of acquired characters was "a deductive argument dependent upon the validity of his claim that a lineage of germ cells was significantly distinct from the soma."[33] This was the origin of what would later be referred to as the "Weismann barrier." The germ cells were partitioned from the somatic cells, passing hereditary material from generation to generation. Any environmental impact on the body, whether through injury or exertion, that did not affect the germ cells would not be passed on to the next generation.[34]

The establishment of the Weismann barrier did not mark the end of the Lamarckians. In 1888 Weismann had conducted a famous—albeit gruesome—experiment on mice. He cut off the tails of the mice and observed that this made no difference to their offspring; the next generation of mice possessed the same tails as their premutilated parents. If the Lamarckians were correct, argued Weismann, the tails of the mice should have eventually disappeared over the generations. The Lamarckians, ironically enough, adapted their claims to counter this evidence. Most now claimed that negative effects on the organism, including injuries, were not likely to be inherited by subsequent generations. They instead turned their attention to the inheritance of positive effects.[35] The acquisition of a new trait through grafting or transplantation counted as a positive effect. Although graft hybrids had been used by Darwin in his defense of pangenesis, however, they did not naturally fit into the Lamarckian framework. Acquired characters had generally been assumed to emerge through the use or disuse of an organ (as with Lamarck's example of the giraffe's neck) or as a direct response to a change in environmental conditions; for example, growing longer hair when placed in a colder climate.[36] Graft hybridization did not fall neatly into either of these categories. It was much more akin to traditional sexual hybridization, which Weismann and his supporters accepted. The only difference was that the hybridization event supposedly occurred via the somatic cells, not the sex cells. If this did occur, the Weismann barrier had been breached. Moreover, the very fact that the body could somehow be involved in heredity implied that the environment it was exposed to might also play a role in inheritance.

When it came to heredity, even the interior of the cell was a contested space. The Weismann barrier confined hereditary material not only to sex cells but specifically to their nuclei. Against this interpretation was the concept of cytoplasmic inheritance, which argues for the heritability of organelles in the cell (other than the nucleus). Cytologists—scientists who study cells—were one of several groups investigating how heredity worked in the early twentieth century. The field of cytology was closely related to embryology, with both disciplines examining how and why egg cells were able to develop complex organisms. Embryologists and their advocates, including the physiologist Jacques Loeb, attempted to take a more holistic view of inheritance and development, which involved speculating on "the existence of factors of heredity located in the cell cytoplasm."[37] Cytoplasmic inheritance pushed the Weismann barrier beyond the cell nucleus and chromosomes, but still restricted heredity to the sex cells.

Cytoplasmic inheritance provided an alternative to a strict interpretation of Weismann's theory well into the twentieth century. Its advocates included many European geneticists who experimented with plant hybridization.[38] In 1909 the German geneticist Erwin Baur showed that pigment-carrying plastids (the small organelles found in plant cells) were passed down through the maternal line of plants. Chloroplasts (where photosynthesis takes place) are a kind of plastid. Other kinds of plastids store energy or the pigments that give plants their color. Baur found that these plastids were inherited (just like the nucleus) across the generations, resulting in such phenomena as plants with distinct patches of green and white in their leaves.[39] Some German biologists used cytoplasmic inheritance to explicitly challenge the "nuclear monopoly" associated with American genetics.[40] Although cytoplasmic inheritance did provide an alternative to the most rigid interpretation of Weismann's results, its adherents were not necessarily open to even more radical forms of heredity like graft hybridization.

It is a sign of the complexity of heredity in the twentieth century that we find Baur acting as a steadfast opponent of graft hybrids and Lamarckism. He was also the originator of the modern chimera hypothesis, which debunked such splendid graft hybrids as the Florentine Bizzaria, and a critic of the Lamarckian Paul Kammerer, a Viennese zoologist who worked on graft hybrid salamanders. An interest in cytoplasmic inheritance did not necessarily lead to the embrace of graft hybridization (which went one step further than cytoplasmic inheritance by implying that somatic cells had a role in heredity). Graft hybrids, however, sometimes provided useful ammunition for those who wished to expand heredity beyond the confines of the Weismann barrier. The twentieth-century triumph of nuclear genetics over cytoplasmic inheritance was in part due to the inability to obtain experimental evidence of the latter and establish the "scientific techniques required to effect major change."[41] Even the briefest appearance of a graft hybrid threatened to reverse this imbalance, providing living proof that heredity was not confined to the sex cells.

With the rediscovery of Mendelian genetics in 1900, a new theory of heredity, free from the inheritance of acquired characters, presented itself for Weismann's scrutiny. At first Weismann was cautious. Experiments on gray and white mice at the laboratory of American geneticist William E. Castle seemed convincing. On the other hand, a 1901 report by two British Mendelians, William Bateson and Edith Rebecca Saunders, seemed to indicate to Weismann that the application of Mendelian genetics was not universal.[42] Initially, then, "for Weismann and the Weismannians Mendel's results seemed interesting but not central to the process of heredity." From this interpretation, Mendelian genetics only increased in importance after 1912, when chromosomes were established as the seat of heredity and reduction division in cells was better understood.[43] Nor were the later views of Weismann entirely clear, as he defended the idea that variation in the germplasm could have an external or developmental cause. This ambiguity "blurred the strong distinction [between heredity and development] that biologists, particularly geneticists, were forging." Weismann and the Weismann barrier were readily adopted by early twentieth-century geneticists in support of their science. However, some Lamarckians would also reinterpret Weismann to support their own views.[44]

Other ambiguities and seeming contradictions swirled around early twentieth-century heredity. In 1920 American botanist Albert Francis Blakeslee pointed out that mutations in plants were not confined to the sex cells. Genetically identical plants could therefore differ in their appearance from one another, thanks to changes in the number or organization of chromosomes in certain parts of their bodies.[45] Blakeslee distinguished chromosomal from genic mutations, attempting to induce the former using radium. His efforts, however, were overshadowed by the experimental success of the geneticist Hermann Joseph Muller, who focused on the role of the gene in heredity and induced mutations using X-rays.[46] In other national contexts, Mendelians and Lamarckians existed side by side. For a brief time, Mendelian genetics was praised by the Soviet Union for "establishing incontestably materialist laws of individual heredity." Yet genetics was believed to be incapable of explaining all facets of evolution, particularly the role of environmental change. Lamarckians, with their focus on the environment, were therefore also accepted. When Kammerer was accused of scientific fraud in the West, he was offered a laboratory in Russia by the Communist Academy's Section of Natural and Exact Sciences in 1925.[47] This period of tolerance came crashing down during the Lysenko affair. As we will see, a strong connection between Mendelian genetics and the Weismann barrier was not a logical necessity. It was, however, adopted by some geneticists to eliminate any hint of the inheritance of acquired characters and reinforce the exclusivity of their science. Heredity was an evolving area of scientific inquiry, the parameters of which had not yet been set.

Mendelian Genetics and Agriculture

During the mid-nineteenth century, a series of hybridization experiments on peas by an Augustinian friar threw up a series of intriguing results. Gregor Mendel's findings can be understood as a kind of algorithmic or mathematical system, which predicted what one should expect when crossing different plants.[48] Mendel began his experiments with garden peas, which have simple inherited traits, easily tracked across the generations. The peas are, for example, either yellow or green, smooth or wrinkled. If you crossed a plant with green and round (gR) peas with another with yellow and wrinkled (Yw) peas, you might end up with a new hybrid bearing yellow and round (YR) peas. In this case, the yellow and round (YR) characters were what Mendel defined as "dominant" factors, while the green and wrinkled (gw)—which did not appear in this first generation—were "recessive," hidden away within. Mendel did not stop there. He then allowed his hybrid pea plants, which in this case were always yellow and round (YR), to self-pollinate. Instead of simply producing more of the same, the offspring of these plants possessed a mix of characters. Some of their peas were yellow and round (YR), some yellow and wrinkled (Yw), some green and round (gR), and the odd one was even green and wrinkled (gw). Mendel had found that both dominant and recessive characters were passed down the generations, with the former masking the latter in the first generation. The combination and recombination of pea plants led to some of the recessive characters emerging in a predictable ratio. With a single character in play, this ratio was approximately three to one in favor of the dominant character. In our example, with two characters in play, the ratio in the second generation was nine to three to three to one (with the dominant yellow and round (YR) again more numerous).[49]

That is Mendel and his insight. By carefully selecting and breeding simple organisms, the Augustinian friar had been able to remove "the baffling clutter, the signal-muffling noise that defeated previous investigators" from the mysteries of heredity.[50] Extracting predictable ratios from the heritable traits of cross-bred pea plants was not only of theoretical interest. Mendel's abbey at Brno, in the modern-day Czech Republic, was not some hermetic retreat. Producing new and stable hybrids was of great interest to farmers in the region, which was particularly famous for its sheep. Mendel's abbot had been active in the world of sheep breeding, while Mendel himself had been trained in horticulture and was a member of the Natural Science Section of the Agricultural Society in Brno.[51] Despite his connections to agricultural interests and other plant hybridizers in Europe, however, Mendel's discoveries famously failed to produce much in the way of scientific or economic impact in his lifetime. In 1900 three academic biologists with interests in plant breeding—Hugo de Vries, Carl Correns, and Erich von Tschermak—independently rediscovered Mendel's laws. The Augustinian friar was subsequently declared the founder of genetics. However, the near-simultaneous claims of uncovering Mendel served the individual interests of the scientists involved, rather than pointing to some historical inevitability that Mendel would eventually be proved right.[52] Key principles in genetics were developed only after the 1900 rediscovery, while de Vries ended up sidelining Mendelian genetics in favor of his own theory of heredity.[53] Regardless of the specific circumstances of who discovered what and when, by the early years of the twentieth century Mendel had followers in the scientific community, some of whom made grandiose claims that the friar's system would revolutionize agriculture.

The extent to which Mendelian genetics could be applied to early twentieth-century agriculture is also controversial.[54] Appeals to the mathematical elegance of Mendel's work likely carried little weight with farmers who, as we will see, were much more interested in practical results than in theoretical explanations. There were good reasons, however, why Mendel's system might be embraced by some. In the United States, the new science of genetics fit with a preexisting trend toward the rationalization of agriculture and an influx of new capital into that sector.[55] The emergence of "genetic rationality," or the bookkeeping involved in recording and organizing data on Mendelian crosses, reflected an emphasis on systematic and rational administration already embraced by industries as diverse as transport and consumer research.[56] Another aspect of Mendel's work that appealed to capitalist thinking was that economically useful traits in plants and animals existed as discrete entities that could be carried between organisms through hybridization and passed down onto future generations. This raised the possibility that living things could be subject to the same intellectual property laws that governed machines and other forms of technology. A 1906 bill placed before the US Congress, for instance, argued that seeds could be patented, as they were mechanisms in the same way a trolley car was.[57] Outside of the United States, growing demands for food and raw materials from urban populations and industry led to the foundation of academic institutions and agricultural stations devoted to plant breeding. Leading facilities were established in Scandinavia, Germany, and Russia by the early twentieth century, at the very moment that Mendelian genetics came into being.[58]

Mendelian-style breeding was not universally welcomed by farmers, however. Sexual hybridization of crop plants, particularly on a large scale, was time-consuming and costly. Control of plant breeding and seeds therefore moved away from farmers and into the hands of private enterprise.[59] Genetics faced a turbulent reception in different national contexts. In France, the "alleged predictability" of Mendelism fell apart when faced with the reality of cereal breeding in the 1900s.[60] A similar problem arose in Britain, where the famed plant breeder Rowland Biffen struggled to apply Mendelian principles to breeding disease-resistant wheat.[61] The worlds of breeding and agriculture were hotly contested spaces, with their inhabitants able to choose from an array of techniques and theories to suit their needs. One of the better-known options was de Vries's mutation theory. In 1901, de Vries, one of the three rediscoverers of Mendel, used the plant *Oenothera lamarckiana* (now *Oenothera glazioviana*) to argue that mutation allowed evolution to occur in sudden leaps and bounds, not slowly and gradually as argued by Darwinians. De Vries visited the United States in 1904, where his mutation theory raised hopes that a shortcut to breeding new types of plants and animals had been found, quelling fears of a growing population outstripping food supply.[62] Unfortunately, de Vries had chosen a plant with highly complex genetics that gave out the false impression of rapid evolution. Excitement over the economic applications of his theory was short-lived, with its popularity beginning to wane by the outbreak of the First World War.[63] As we will explore, the graft hybrid represented another contender for agricultural improvement.

The science of genetics did not stand still across the twentieth century. New findings altered the discipline and informed its adherents' attitudes to graft hybridization. One of these findings came about through the interaction of an American geneticist, Thomas Hunt Morgan, with a new model organism: the fruit fly (*Drosophila melanogaster*). The fly had many advantages for laboratory study. It had a simple genome, was easy to breed, and had a short lifespan, thereby cutting down on the time and cost usually involved in genetic research.[64] Morgan and his students observed the Mendelian system at work in fruit flies and went on to theorize that the chromosome was the seat of heredity. Beginning in 1910, they found that certain characters—such as eye color—were linked to patterns of sex chromosome inheritance in flies. When Morgan teamed up with Frans Alfons Janssens, a Belgian cytologist, they hypothesized that Mendelian characters were carried along the chromosomes, crossing between them during cell division with a frequency relative to their position. Morgan used these insights to argue that genes were the units underpinning heredity. He and his supporters subsequently "deemphasized hereditary phenomena that could not be explained by their theory."[65] This assertion caused some embryologists to protest "the exclusive role of the nuclear gene in heredity," arguing that Mendelian genetics accounted for only relatively trivial characteristics within a species. The cytoplasm, they claimed, played a more important role in fundamental evolutionary change.[66]

Another important aspect of twentieth-century genetics was eugenics, the troubled application of heredity principles to humanity. Many of those involved in the various controversies over the existence of graft hybrids were enthusiastic eugenicists, including Bateson, Baur, Castle, and their fellow geneticist Charles Davenport. Some graft hybridizers were also eugenicists. Kammerer advocated a kind of Lamarckian eugenics, arguing that environment and education could heal and improve humanity.[67] The actual incorporation of eugenics into scientific discussions of graft hybridization was rare, but not unheard of. An exchange in the American *Journal of Heredity* in 1927 included a contribution by the geneticist Robert C. Cook, who argued that references to cross-species grafting in the plays of Shakespeare indicated that they were actually authored by the natural philosopher Francis Bacon. Cook suggested that it would be near-miraculous if "the random Stratford boy, abandoning his wife and children at twenty" could have produced such masterpieces. Under the eugenic framework, genius and scientific knowledge were more likely to be the hallmarks of a morally upstanding member of society from an esteemed family. "The authorship of the plays by a person with Bacon's breadth of interest and literary endowments," argued Cook, "is much more explicable biologically."[68] Overall, however, graft hybrids were of more interest for their apparent ability to breach both the Weismann barrier and the barriers between species. The conflict between geneticists and graft hybridizers would later take on a political flavor with the outbreak of the Cold War.

Grafting in the Cold War

One of the darker episodes in the history of science took place in the Soviet Union with the rise of Trofim Lysenko, a peasant-farmer-turned-agronomist. His career had begun amid the 1927 collectivization campaign in the Soviet Union, when Joseph Stalin persecuted those deemed "kulaks" (wealthy peasants) and instigated a famine in Ukraine. Soviet propaganda of this period depicted Lysenko as a "barefoot scientist" whose practical, almost rustic, skills had earned the gratitude of the people and the praise of agronomic experts.[69] By 1933, Lysenko had taken control of the Institute of Plant Breeding and Genetics in Odessa, despite a lack of any formal scientific training. Here he promoted vernalization—a technique involving the exposure of growing seeds to cold in order to speed up their progression to the point of flowering—of major crops such as wheat and cotton. Plant physiologists in the Soviet Union cautiously accepted some aspects of Lysenko's work, which found a more welcome reception among the Communist Party bureaucracy. Vernalization, however, did not live up to Lysenko's promise and was quietly dropped.[70] Lysenko's early forays into heredity and genetics had been roundly criticized by both plant breeders and academic biologists. By 1935, Lysenko's political standing had raced ahead of his achievements. He was lauded by newspapers, supported by Stalin, and a member of the prestigious V. I. Lenin All-Union Academy of Agricultural Sciences, becoming its director in 1938.

Lysenko's final triumph over the entirety of Soviet biology came in an infamous meeting of the Academy of Agricultural Sciences in 1948. Lysenko denounced classical genetics, claiming that Weismannism and Mendelism-Morganism (the latter term referring to the idea that chromosomes contain heredity information) had "been primarily directed against the materialist foundations of Darwin's theory of evolution."[71] In short, he was throwing away the whole of genetics and chromosome theory as a harmful bourgeois science, antithetical to the values of the Soviet Union. Protesting this attack would have been unwise. By the time of the 1948 meeting, Lysenko's primary opponents were dead or had been forced aside. The botanist and agronomist Nikolai Vavilov, the former director of the Academy of Agricultural Sciences, had died in prison in 1943.[72] Muller, an American, had fled the Soviet Union some years earlier. In a misguided attempt to win Stalin over to the geneticist camp, Muller had sent the Russian dictator a copy of his book, *Out of the Night*, which promoted eugenics. Stalin was displeased and Muller fled.[73] In case there was any doubt as to where the dictator's sympathies lay, Stalin carefully proofread and edited Lysenko's 1948 address.[74] With such powerful political backing, Lysenko was able to lay out his vision for biology, which included a form of Lamarckism and an emphasis on cooperation between organisms, without much in the way of evidence that any of his ideas worked.

Here is where the graft hybrid came in. Graft hybridization had maintained a respectable presence in biology well into the 1930s. This history meant that Lysenko's claim that new plants could be produced using grafting was one of his few ideas with any credibility beyond the Soviet Union. In 1945, Lysenko displayed some tomato graft hybrids at a lecture to mark the 220th anniversary of the Russian Academy of Sciences. Julian Huxley, evolutionary biologist and science writer, was in attendance. A British colleague suggested that fraud was afoot, with the plants displayed by Lysenko probably obtained from existing varieties, not by grafting. Huxley also noticed that the tomatoes displayed at the lecture were wax models.[75] At his 1948 address to the Academy of Agricultural Sciences, Lysenko displayed more graft hybrids. He declared that their existence breached the Weismann barrier and lay beyond the explanatory power of classical genetics. He was not, however, open to discussing how or why graft hybridization occurred. In 1949, Jean Brachet, a Belgian embryologist and member of Belgium's Communist Party, visited Lysenko. Brachet suggested that graft hybrids might be the result of "self-replicating virus-like genetic particles in the cytoplasm" of plants, which could reach out and invade the rest of the body. He proposed an experiment to reveal whether this was the case, suggesting that a membrane be inserted between two grafted plants that would block the passage of these hypothetical particles. Lysenko had no interest in experiments conducted only for scientific curiosity, which he regarded as a symptom of capitalist excess. Brachet returned home and denounced Lysenkoism.[76]

Clearly, then, Lysenko did not draw upon the experimentally minded graft hybridizers of Western Europe and North America for inspiration. His vision for biology may have been informed by a much older tradition, with its roots in the nineteenth-century acclimatization movement. Acclimatization—the theory that plants and animals could adapt to new environments over time—was embraced by prominent Russian thinkers associated with the "Westernization movement" of the 1840s. The ambition to control the evolution of species through the environment, expressed by bodies such as the Moscow Agricultural Society and the Imperial Russian Society for the Acclimatization of Animals and Plants, would become an integral part of Lysenkoism.[77] Neo-Lamarckism was also prevalent in the Russian life sciences. Ivan Pavlov, famed for his behavioral experiments with dogs, was a believer in the inheritance of acquired characters and suggested that even behavioral reflexes (such as dogs salivating at the sound of dinner bells) could be ingrained in the organism over generations.[78] Kammerer, a Lamarckian, was idolized as a hero in the early Soviet Union. The 1928 Soviet-German movie *Salamandra*, which was produced after claims of scientific fraud and

Kammerer's suicide, depicts Kammerer as the victim of a conspiracy who is eventually saved by the Soviet Union.[79]

Lysenko claimed that one of his most important influences was Ivan Vladimirovich Michurin, a Russian horticulturalist. During the early years of the twentieth century, Michurin had achieved a level of fame as a self-taught plant breeder, producing hybrid fruit trees that could replace imported varieties. He also believed in the existence of graft hybrids, coming up with his own "mentor" method of producing them. This involved grafting young shoots onto old stock, in the belief that the more mature plant would exert a greater influence over the developing scions. In later life, Michurin was lionized by the Soviet state, which portrayed him as a patriotic hero fending off offers from American capitalists to buy out his research.[80] Lysenko would adopt Michurin's beliefs regarding graft hybridization wholesale. Michurin, however, never denied the reality or applicability of Mendelian genetics. We will see how Western commentators realized that Michurin had been misrepresented by Lysenko, and explore how "Michurinism" became both a substitute term for Lysenkoism in the Soviet Union and a label with which one could express support for graft hybridization in the West without mentioning Lysenko. As for Lysenko's biology, we are forced to agree that "where he was right, he was not original; where he was original, he was not right."[81]

The Lysenko affair, as the Soviet attack on genetics is sometimes called, had global repercussions. Renewed contact with Soviet scientists after the Second World War gave their Western colleagues the impression that Lysenko's position was not unassailable. Geneticists in Britain and the United States organized anti-Lysenko campaigns. Among their number were the exiled Muller and Huxley, who had seen Lysenko's wax models of graft hybrids firsthand. The outbreak of the Cold War, however, would cause these international links to become a liability to Soviet geneticists.[82] When Lysenko asserted his scientific and political dominance in his 1948 lecture, Lysenkoism began to be labeled as "pseudoscience" in the United States.[83] One casualty of the growing Cold War divide in biology was research into chromosomal mutations. In 1927, Muller had bombarded the sperm of *Drosophila* fruit flies with X-rays, creating alterations to their chromosomes that could be passed down through three or four generations.[84] In the United States this result was taken as the result of genetic mutations. In the Soviet Union it was interpreted as an example of the influence of the environment on inheritance.[85] Lysenko was not inclined to alter the organism through chemicals or radiation. Although he did not deny the effects of such methods, he viewed them as a kind of "poison" that "can only rarely and only fortuitously lead to results useful for agriculture."[86] During the 1940s and 1950s American biologists also downplayed the importance of chromosomal mutations. Heredity was simplified to the level of the gene, while the mere fact that Soviet biologists had shown an interest in chromosomal mutations was enough to dissuade their American counterparts from following suit. Now, "heredity in the West was increasingly defined, refined, and constrained in opposition to Lysenkoist interpretations."[87]

In other national contexts, a visceral rejection of all things Lysenko did not occur. One example is that of postwar Japanese genetics. When information on Lysenko's experiments was circulated among Japanese geneticists during the 1940s and 1950s, some were intrigued by Lysenko's graft hybrids. Hitoshi Kihara, an internationally prominent expert in wheat genetics at Kyoto University, even "suggested some alternative possibilities to interpret the graft hybrid from the viewpoint of orthodox genetics," a stance that was also adopted by British biologists with sympathies for Soviet biology.[88] Kihara was far more critical of other aspects of Lysenkoism. In 1953 he and geneticist Karl Sax coauthored an article in which they mocked the Lysenkoist claim to have transformed wheat into oats or rye by planting it in different environments. "By inference," they wrote, "we might assume that under suitable conditions, perhaps by proper housing or diet, the Soviet scientist will be able to convert Orang-outangs into humans or vice versa."[89] Unlike their American counterparts, Japanese geneticists made space for the role of the cytoplasm in heredity. The appearance of the inheritance of acquired characters in Lysenkoism did not unduly worry Japanese researchers either. Kihara noted that when some bacteriologists in Japan called themselves Lysenkoists, they only meant that they worked on adaptive mutations or the inheritance of acquired characters.[90]

In addition to the denouncement of Lysenko, the Cold War also saw the crowning of Mendel as the "father of genetics." In 1950 the Genetics Society of America celebrated the fiftieth anniversary of the rediscovery of Mendel, generating widespread radio and newspaper coverage. The society had held back from speaking out against Lysenkoism, but now found the perfect opportunity to use the celebrations "to present a positive, dignified, and powerful alternative."[91] Many of the participants in a scientific panel assembled by the society, including Huxley, were aware that the general public could become lost amid the subtle distinctions between neo-Lamarckism, cytoplasmic inheritance, and gene expression. Their solution was to present

Mendel as the common ancestor for all the varied forms of modern genetics. Not every biologist was willing to accept this elevation of Mendelism. The geneticist and future Nobel Prize winner Barbara McClintock stayed away on the grounds that the Genetics Society's anniversary event was "a celebration of the triumph of classical genetics."[92] This period also saw the association of Mendelian genetics with the development of hybrid maize in the United States. During the Cold War, this event would be promoted as a triumph of Western plant breeding over the agricultural failures of Lysenkoism.[93]

The Cold War did strange things to the graft hybrid. Graft hybridization maintained a small, yet respectable, following in Western biology into the 1930s. The rise of Lysenko shook up this cozy situation. The Cold War thrust the graft hybrid back into the scientific spotlight and subjected it to fresh scrutiny, decades after some of the most intense exchanges over their existence had passed. Unfortunately for defenders of the concept, graft hybridization was now associated with all the unpleasantness of Lysenko and the Stalinist regime. As we will explore, some defenders of Soviet science, notably Haldane, halfheartedly pointed to graft hybridization as evidence that Lysenko's theories had some validity. McLaren turned this approach on its head, using the relative strength of graft hybrids as a scientific concept to promote her version of Marxist biology. The Cold War division of biology had a lasting impact. We will see how practitioners of cell fusion were wary of comparisons to graft hybridization. Although the two techniques had their similarities, the graft hybrid had become a tainted idea. The twentieth century would pass before its existence was again debated in Western science.

Structure and Concluding Remarks

Before we begin our journey into the world of the graft hybrid, a few words of caution. Heredity in the twentieth century is a complex affair. Each of the actors we encounter in this book was a complicated character, and many of them held what we would today consider ambiguous or contradictory beliefs. William Bateson, the British Mendelian who attempted to interpret the development of graft hybrids, did not believe in chromosome theory.[94] His colleague the German geneticist Erwin Baur attacked various graft hybridizers over the course of his long career in biology. Baur, who pursued research on mutation and recognized the complex relationship between Mendelian genetics and the environment, was no ordinary Mendelian. His full-throated defense of the field was directed "against those who tried to limit Mendelian validity," such as the graft hybridizers.[95] The graft hybridizers themselves had complex beliefs that did not necessarily clash with genetics. Charles Claude Guthrie thought that his graft hybrid chickens might simply represent an extra hereditary mechanism. Paul Kammerer and Ivan Vladimirovich Michurin, both graft hybridizers, also accepted the validity of Mendelian genetics. A similar level of complexity haunts our efforts to define what a graft hybrid was at any given time. I have attempted to stick with the spirit of Darwin's definition when referring to "graft hybrids," as it was usually the case that a plant or animal hybrid created through grafting or transplantation was called a graft hybrid. Nevertheless, there were exceptions to this rule.[96]

This book consists of six chapters, ordered chronologically, which trace the graft hybrid throughout the twentieth century and its revival in the twenty-first. These chapters do not constitute an exhaustive account of every twentieth-century graft hybridizer, but do focus on important debates, collaborations, and exchanges between notable players in the field.[97] Chapter 1, "A Poultry Affair," jumps straight into an early twentieth-century graft hybrid controversy, examining the uptake of Mendelian genetics in the United States, and how and why Guthrie, an American physiologist, came to believe that he had created graft hybrid chickens. A series of indecisive back-and-forth experiments between Guthrie and the geneticist William E. Castle sets the scene for the debate between graft hybridizers and geneticists for the rest of the century. In chapter 2, "Rise of the Chimera," I consider what was defined as a graft hybrid in more detail, focusing on the scientific reception of a tomato-nightshade hybrid created by the German botanist Hans Winkler. In chapter 3 I explore how scientific belief in graft hybridization persisted into the interwar period, following the graft hybrid salamanders of Paul Kammerer and efforts by the British botanists Frederick Ernest Weiss and William Neilson Jones to locate graft hybrids and account for their origins.

After the Second World War and Lysenko's attacks on geneticists in the Soviet Union, the graft hybrid became entangled in the wider ideological clash of the Cold War. In chapter 4 I explore this tension through a series of encounters between British scientists and plant breeders with their counterparts in the Soviet Bloc. In chapter 5 we encounter the British zoologist Anne McLaren, whose politics and frustration with the limits of Mendelian genetics led her on a global search for graft hybrids. Over the course of her career, McLaren would encounter graft hybrid poultry in Hungary, tomatoes in Yugoslavia, eggplants in China, and peppers in Japan. In chapter 6 I describe the relationship between graft hybridization and the field of somatic hybridization (a form of cell fusion) in the mid-to-late twentieth century. Unlike the graft hybrid, cell fusion was once celebrated in the West as the future of biotechnology. Its practitioners attempted to distance themselves and their science from the ideologically charged (but closely related) technique of graft hybridization. I conclude with the contemporary revival of the graft hybrid in the twenty-first century. Although its connotations with Lysenkoism remain problematic, grafting is recognized as a means by which one plant can transfer genes to another. Despite a century of controversy, the graft hybrid has now been incorporated into modern biotechnology.

Before we launch into the history of the graft hybrid, there is one final matter to clear up. What was the true nature of the famous Florentine Bizzaria we encountered at the start of this chapter? Shortly before the outbreak of the First World War, some botanists theorized that plants like the Bizzaria were not true hybrids. They were instead chimeras, two genetically distinct organisms intertwined in a single body.[98] Though named after the fire-breathing monster of Greek mythology, real-world chimeras are quite common. A mutation in the pink flowers of a peach tree, for example, can give rise to genetically altered cells, leaving white patches or flecks in the flower where the new cells have grown.[99] Chimerism can also happen to humans. Cellular traces of a fetus can persist in the body of the mother for years after pregnancy, or two zygotes can occasionally fuse together. More commonly, cells with a different genetic code are introduced to our bodies via medical interventions, such as in organ and bone marrow transplants.[100] The Florentine Bizzaria, however, was more dramatic in its appearance than these examples. Not only was it composed of different species but its cellular tissues of citron and orange were seemingly inseparably mingled.

During the winter of 1922 to 1923, Tyôzaburô Tanaka, a member of the Phytotechnical Institute at the Miyazaki College of Agriculture in Japan, traveled to Italy. Tanaka visited the Botanical Institute of the University of Florence and then the Giardino Botanici Hanbury of La Mortola, where he found a specimen of the Bizzaria. When he examined the plant, he found that it consisted of a citron core surrounded by an external skin of sour orange. Tanaka theorized that the Bizzaria was the result of different elements battling for space in the body of a single plant. He could explain the appearance of the plant using the chimera hypothesis, without reference to hybridization. "Critical study of this much discussed graft-hybrid," concluded Tanaka, "thus brings us to the conclusion that this is a clear case of periclinal chimera."[101] A periclinal chimera refers to a type of chimera in which the genetically different cells occupy different layers of tissue—in the case of the Bizzaria, leading to a plant that is largely citron, but with a skin (or bark, in this case) made of cells from an orange tree. Despite its fantastical appearance, then, the Bizzaria was far from the most intriguing or compelling example of a graft hybrid. It is to these organisms we now turn.

1

A Poultry Affair

——✑

IN 1903, CHARLES CLAUDE GUTHRIE, A TWENTY-THREE-YEAR-OLD postgraduate student at the University of Chicago, received an exciting offer from one of America's foremost physiologists. Elias P. Lyon was popular among his peers, regarded as a superb teacher and one of the greatest scientists in the entire United States.[1] Lyon was looking for someone to assist him in a series of experiments, with the goal of seeing whether ovaries could be transplanted from one animal into another and continue to function. To Guthrie, who originally hailed from a Missouri family farm, the chance to work alongside Lyon must have seemed like the opportunity of a lifetime. But their surgical experiments on chickens did not go according to plan. Guthrie and Lyon failed in their first attempt during February 1904. Lyon then left Chicago to take up the directorship of another laboratory, leaving Guthrie behind. Guthrie's second attempt at transplantation in August that year succeeded, but his hens were kept in "inappropriate quarters," so that "no eggs were laid, even by the controls."[2]

Guthrie, despite two failed attempts and the loss of his mentor, did not abandon his experiments on chickens. On August 25, 1906, he performed a third set of operations, surgically transplanting ovaries between two varieties of chickens. These hens survived and were sent to recuperate at a poultry yard in Columbia, Missouri, before moving to Guthrie's new laboratory

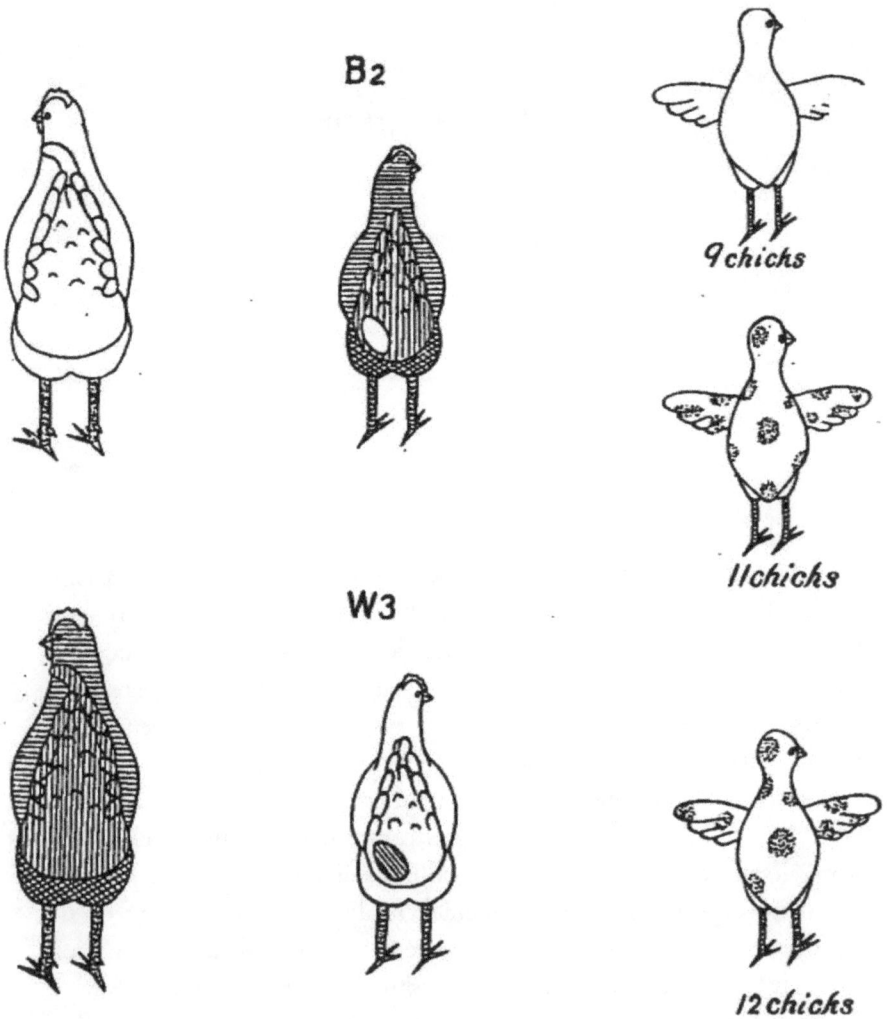

FIGURE 2. Transplantation in chickens. Charles Claude Guthrie's experiments on chickens involved transplanting ovaries between breeds of different colors. When the offspring of these hens showed a mix of colors, Guthrie claimed that some unknown heredity mechanism was at work. His experiments provoked a backlash from Mendelian geneticists. C. C. Guthrie, "On Graft Hybrids," *Journal of Heredity*, o.s., 6, no. 1 (1911): 363. Reproduced by permission from Oxford University Press.

at Washington University Medical School, where he had been appointed professor of physiology and pharmacology. Guthrie now began breeding his chickens to ascertain whether the implanted ovaries still functioned. When he did so, he found something unexpected. He had selected two lines of purebred Single-Comb Black Leghorn and Single-Comb White Leghorn chickens—one with entirely black feathers, one with entirely white—for his transplantation experiment. When a black-feathered male was crossed with a white-feathered female with transplanted ovaries from a black-feathered female, the offspring were a mixture of black and white. The same combination of colors occurred in the feathers of offspring born of a union between a white-feathered male and a black-feathered female with ovaries from a white-feathered female (fig. 2).

When Gregor Mendel's work was rediscovered in 1900, some of its more enthusiastic supporters interpreted hereditary characters in an atomistic fashion, arguing that they could be removed from or inserted into agricultural plants and animals at will. Guthrie had seemingly achieved the same thing, combining the traits of two different animals in their offspring. Yet he did this without the use of a Mendelian breeding program, in a way that seemingly breached the "Weismann barrier" between inherited sex cells and the cast-off cells of the body. Organ transplantation had altered heredity in a way that should have been impossible according to August Weismann's claims. Guthrie is today remembered—if at all—for his medical work alongside Alexis Carrel and his exclusion from the 1912 Nobel Prize in Physiology and Medicine.[3] A French biologist and eugenicist, Carrel was undoubtedly a skilled surgeon and a pioneer of tissue culture.[4] He was also a publicist with grandiose ambitions to achieve immortality through transplantation.[5] Like Carrel, Guthrie was admired for his technical skill but had his larger visions for biology dismissed.

During his physiological experiments on chickens, Guthrie created what he would later describe as "graft hybrids." Prominent Mendelian geneticists in the United States, including William E. Castle and Charles Davenport, disagreed with Guthrie's findings. But before we explore this debate, let us for now turn to William Bateson, the British scientist who coined the term *genetics* in 1905, his embrace of Mendel, and the reception of his newfound science of genetics in the United States.

Mendel's Bulldog in America

In the spring of 1886 William Bateson, a young biologist, set out on an expedition from Cambridge, England, to Siberia. One of his goals was to find empirical evidence in favor of the Darwinian orthodoxy of his time, which held that plants and animals were strongly influenced by their environment through natural selection. Another was to test the claim of a Russian neo-Lamarckian that brine shrimps could be transformed into a new species in only a few generations by changes in the salinity of their aquatic habitat.[6] Bateson endured months of hard travel and illness, leaving broken and worn-out scientific equipment in his wake. Upon his arrival among the lakes of Siberia, however, his optimism, at least, remained intact. In a letter written from the town of Pavlodar, he told his family that he was delighted to have found shallow lakes ideal for his work. "My plan," he explained, "is to wade in, tow a few yards until there are enough crustacea to make a solid lump at the bottom of [the] tow-net; this lump I tip into a bottle and pour in sublimate." The only obstacle to Bateson's plan came when he attempted to measure the density of his captured marine organisms, as "my beautiful long glass beaker is broken, so I am compelled to use an old jackboot for this purpose, which, being a quite unscientific instrument, had better not be repeated."[7]

Despite such difficulties, Bateson threw himself into his work, riding for miles across the Siberian steppes from lake to lake. By skipping meals, he wrote (to what was probably an increasingly anxious family), he managed to work for up to twelve hours a day. Bateson attached great significance to the outcome of his fieldwork:

> I shall be more intelligible if I say that I shall undoubtedly have enough facts to show me whether these beasts do or do not show variations proportional to the salinity of their habitats. If they do shew such variations, I shall be supremely happy, and it will serve as the basis for any number of life-works; and if they don't, that will be worth knowing too, and I shan't feel that I have thrown this time away, though that will mean chucking overboard "cherished" convictions and looking out for a new basis for all those life-works.[8]

Upon his return to Cambridge in 1887, however, Bateson felt unable to support either the neo-Lamarckians or the thesis that the outward appearance of organisms could be linked with their environment. One consequence of his rebellion was that he fell out with the morphologists who dominated biology at the University of Cambridge.[9] Another was that his attitude toward neo-Lamarckism hardened considerably. Several decades later, his son Gregory Bateson recalled that his father "was certainly not a nice man whenever the inheritance of acquired characteristics was mentioned. When this happened the coffee cups rattled on the table." The younger Bateson related that the elder had gone "to the [Siberian] steppe in order to prove the inheritance of acquired characteristics" and that the failure of this project had "something to do with his later attitude . . . he regarded Lamarckism as a tabooed pot of jam to which he was not allowed to reach."[10] By the close of the nineteenth century, William Bateson was a very different creature from the young man who had waded through Siberian lakes. He was not only on the lookout for a new basis on which to conduct his scientific work but actively hostile to the notion that changes to an individual's body induced by the environment influenced heredity.

In October of 1902 a train carrying Bateson, now a proud Mendelian, drew into Ithaca, New York. Bateson had arrived in America on a lecture tour to promote the study and application of Mendelian principles. Liberty Hyde Bailey, professor of horticulture at Cornell University and an admirer of the new Mendelian doctrine, had already roused excitement among his compatriots by declaring Bateson's latest book, *Mendel's Principles of Heredity: A Defence*, indispensable to the plant breeder. His support paid dividends. "At the train yesterday," wrote Bateson, "many of the party arrived with their 'Mendel's Principles' in their hands! It has been 'Mendel, Mendel all the way,' and I think a boom is beginning at last."[11] Yet when Bateson had returned from his disappointing Siberian fieldwork in 1887, he had been something of a maverick among his fellow biologists in Cambridge. What had changed to ensure that he would be greeted by parties of fervent supporters when he crossed the Atlantic some five years later?

When Bateson abandoned the Darwinian orthodoxy of his day, which insisted that there were a series of gradually changing characteristics between different species, he had turned his attention to experiments in hybridization. When Bateson aired his controversial views, he managed to simultaneously irritate the Cambridge establishment and spark a long-running feud with his former friend W. F. R. Weldon. On July 11, 1899, Bateson was invited to London by the Council of the Royal Horticultural Society, which was holding a special conference on the hybridization of plants. Bateson rose to the occasion, announcing that a new form of natural history could be created through the study of hybrids to define the boundaries between species. Even better, all that was needed for this new natural history was, "leisure, accuracy, and a garden of moderate extent" to produce scientific results of great importance.[12] Little did Bateson know that a certain Augustinian friar, Gregor Mendel, had already carried out the large-scale hybridization experiments that he now proposed to the Royal Horticultural Society.

Another attendee at the 1899 conference was a Dutch botanist by the name of Hugo de Vries, who would soon become known as one of the re-discoverers of Mendel's 1866 paper on pea hybridization. Bateson himself recognized the significance of Mendel's findings after reading one of de Vries's papers on the train from Cambridge to London during the summer of 1900. He quickly appointed himself as Mendel's defender and advocate with all the zeal of a new convert, hence his popularity among Mendelians at the Ithaca train station.[13] Shortly after his station reception, Bateson told the New York Horticultural Congress that, thanks to Mendel's insights, breeders now knew that the organism was a collection of traits: "You can take out greenness and put in yellowness; you can take out hairiness and put in smoothness; you can take out tallness and put in dwarfness."[14] This view of heredity, which depicted the characters of plants and animals as atomistic traits to be removed and added at will, would clearly have been attractive to the horticulturalists of New York. Rather than relying on guesswork and randomness, they would be able to effectively design the plants they wanted.

Bateson went on to receive a friendly welcome from some of the great minds of American biology. In 1907 he was once again invited back to the United States to deliver the prestigious Silliman Lecture series at Yale University. True to form, Bateson chose once again to court controversy. Perhaps stung by the memory of his fruitless Siberian labors, he comprehensively dismissed all the "older evidence for the inheritance of adaptative changes," insisting that the experiments of Weismann had debunked the role of environmentally induced changes to the body in heredity. Only "violent" shocks, such as poisoning or starvation, to the reproductive cells were sufficient to induce some kind of adaptation to the environment, declared Bateson.[15] With the old doctrine cast aside, the somatic cells of the body would have little or no role in the new Mendelian account of heredity.

Bateson's arguments were met with great acclaim. Yandell Henderson, professor of physiology at Yale University, praised his lectures as "full of interesting ideas and facts which entertained the laity to a degree no other Silliman Lecturer has been able to do."[16] When Bateson's Silliman Lectures were published in 1913 as *Problems of Genetics*, an anonymous reviewer in *Nature* painted a very similar picture of Bateson's appeal:

The task which Mr. Bateson, as Silliman lecturer, set before himself was a discussion of some of the wider problems of biology in the light of knowledge acquired by Mendelian methods of analysis, and the reader who would get most from the book should have some acquaintance with the phenomena of heredity recently brought to light by the method of experiment. To one with even an elementary knowledge of these phenomena, Mr. Bateson's book cannot fail to prove of absorbing interest. For he has the rare gift of infusing something of strangeness into the commonplace, and in his hands the seemingly familiar takes on an aspect of remoteness which once again provokes curiosity.[17]

If we take the comments of Bateson's colleagues and reviewers at face value, then it would appear that he had managed to capture the imagination of plant and animal breeders across the United States. Bateson had a very different impression. "So oppressive," complained Bateson in his *Problems of Genetics*, were the exaggerated wonders of adaptation and ubiquity of anthropomorphic reasoning, that the association of heredity and the transmission of acquired characters was "a preconception still almost universal among the laity."[18]

Many of those who attended Bateson's lectures and subsequently engaged in discussions about the nature of heredity were not representative of the ordinary American farmer. They were instead professional breeders, or trained biologists conducting agricultural science at experimental stations and land-grant universities.[19] This caveat aside, there was much within Bateson's interpretation of Mendel to commend his theories more broadly to farmers and breeders across the United States. For one, Mendel himself had conducted his experiments in plant hybridization with the eye of a plant breeder, amid an agricultural landscape. Moravia was home to an important wool industry and the meetings of its Sheep Breeders' Society were regularly attended by Mendel's mentor, Abbot Cyril Napp. Strangely enough, the bureaucratic nature of the Mendelian system also appealed to early twentieth-century Americans. Detailed records and statistics were quickly becoming an integral part of American industry, business, and transport. The meticulous collection and organization of data required for Mendelian genetics gave rise to a kind of "genetic rationality," which resulted in the hope that agriculture could finally become efficient and predictable.[20]

But did these hopes of control and predictability over the breeding of plants and animals translate into practical gains? Or did American breeders,

as Bateson feared, cling to the old beliefs in the inheritance of acquired characters? Some insight into these questions can be gleaned from the American
Breeders' Association, founded in 1903. Its members were drawn from the
worlds of commercial breeding, academia, experimental stations, and the US
Department of Agriculture, producing numerous reports in the association's
American Breeders' Magazine (later the *Journal of Heredity*).[21] Early entries in
the association's magazine regarding the practical application of Mendelian
theory can best be described as tentative. One of the first contributors was
William Jasper Spillman, an agricultural scientist who had spotted Mendelian ratios during his work with wheat varieties in eastern Washington.
By this time an employee of the US Department of Agriculture, Spillman
proposed that Mendel's laws could, theoretically at least, be applied to cattle breeding. After all, "Bateson has shown that [Mendel's laws] appl[y] to
many characters in poultry," while the Harvard-based Mendelian William
E. Castle had demonstrated that Mendel's laws were "applicable to various
characters of guinea-pigs, mice and rabbits."[22] Spillman went on to lay out
a scheme for removing the horns of cattle through repeated hybrid crosses
and backcrosses.

American advocates of Mendel, including Castle and Davenport, were
much more optimistic in their assertions that Mendelian laws could be
applied to agriculture. Overall, the number of references to Mendel in the
association's magazine did increase over time, with the majority of contributors who cited Mendel reporting that their crosses conformed to Mendelian
ratios.[23] But was Mendelian genetics actually helping the everyday business
of plant and animal breeding, or was it simply an interesting theoretical
concept? Carl S. Scofield, of the US Department of Agriculture, leaned toward the latter. In a 1905 paper he urged his countrymen to keep detailed
lists and scorecards of the characters they used to judge the quality of new
plant varieties. By doing so, he argued, crop plants would ultimately be more
likely to remain true to type and rogue plants could be more easily eliminated. Although Scofield's recording and scorekeeping appeared superficially
Mendelian, however, he saw the data gathered by such methods as means to
allow "such theories as those of de Vries and Mendel [to] be given wider and
more accurate tests."[24] The outward appearance of Mendelian breeding did
not necessarily equate to breeders' acceptance of its principles.

Of almost equal importance to the uptake of Mendelian principles, at
least in the view of Bateson, was the destruction of their antithesis: the inheritance of acquired characters. Yet belief persisted in the heritable influence

of the environment on the organism, particularly among American horticulturalists. For growers and breeders of fruit trees, which varied considerably
in their genetic makeup from generation to generation and took many years
to reach maturity, Mendel's laws were not as relevant as elsewhere. As late as
1911, Burt C. Buffum, professor of agriculture at the University of Wyoming,
still clung to the idea that fruit trees could be radically altered by changes of
climate or environment. He told the story of how an orchardist in northern
Colorado purchased some apple trees from a commercial nursery that were
so altered by their short spell in his mountainous orchard that the nursery
could no longer recognize the fruit of their own trees. "I cite these instances of changes by environment," explained Buffum, "as an indication of the
somewhat unique position in which the plant breeder of the mountain region
finds himself."²⁵ Climate and altitude were not the only factors that breeders
had traditionally used to induce changes to their crops, however.

The creation of hybrids through grafting, another phenomenon excluded
from Bateson's strict interpretation of Mendelian genetics, was also present
among the pages of the American Breeders' Association's publications. Luther Burbank, the popular "wizard" of American plant breeding, commented
on the nature of grafted plants in 1905, hinting that graft hybrids were
a genuine possibility: "Grafting or budding may be called a bio-mechani-
co-chemical combination. While crossing by seed is more of a bio-chemical
union, yet this last union is often more truly mechanical than chemical as in
the case of a mosaic union, which is not unusual when the cross is too abrupt.
In fact, every gradation from a purely mechanical union to one of perfect
chemical blend is a common every-day occurrence with those who have carried out field experiments on a broad and comprehensive scale."²⁶ Burbank
was for once not unusual in prioritizing messy experience amid fields and
orchards over the ideological divide between Mendelians and Lamarckians.
In his role as director of the South Dakota Agricultural Experiment Station,
Danish-American horticulturalist Niels Ebbesen Hansen felt no need to set
different explanatory systems of heredity phenomena against one another.
In a 1906 report on the breeding of fruit trees, he proudly proclaimed, "By
far the most extended experiment on record in the making of graft-hybrids of the apple has been undertaken" in South Dakota, and "we await the
fruiting of the resulting plants with interest."²⁷ Hansen used the remainder
of his report to make some revelatory remarks. He first commented on the
ability of breeders to control the evolution of new plant species, suggesting
that such work was "more like the labor of an inventor of machinery in his

workshop than that of an observer of an ever changing panorama." Hansen's opinion was not informed by a newfound hope in the science of genetics but by his own experience "first hand with many thousand seedlings of native and cultivated fruits and plants." Hansen then portrayed the plant breeder as riding "his automobile on the highway of evolution," with Mendel's laws representing just one of its wheels.[28]

What, then, was the reigning paradigm among American plant and animal breeders by the close of the first decade of the twentieth century? Had Bateson's interpretation of Mendel triumphed or, as he feared, did a belief in the inheritance of acquired characters still persist among the so-called laity? The exchanges of members of the American Breeders' Association can tell us only so much. What is clear is that Mendelian principles and techniques did gradually rise in popularity and demonstrate some practical applications. For instance, Mark A. Carleton, a botanist at the US Department of Agriculture, remarked in 1906 that the discussion of "Mendel's law and kindred topics" had encouraged plant breeders to "now see, as we did not at first, the importance from a statistical standpoint of securing as great a number of the progeny as possible in the case of each hybrid."[29] In this instance, Mendel's laws had at least encouraged breeders to produce and save potentially valuable hybrids, which might otherwise have been ignored or lost.

Yet the triumph of Mendel in American agriculture was not complete. In horticulture and pomology, where Mendel's laws were less applicable, belief in the importance of the environment and the possibility of graft hybrids persisted. This did not mean that Mendel was rejected outright. In fact, breeders were far more interested in what techniques worked than why they did, or were unwilling to exclude particular techniques from their arsenal simply because they did not fit within a particular theory of heredity. "Mendel's law," explained Carleton in 1906, "like certain machines, is a good thing in grain breeding so long as it works."[30] Although he had praised Mendelian genetics and applied it to his own experiments in grain breeding, Carleton noted that the theory did not work everywhere. Buffum, who would argue in favor of the influence of the environment on plant varieties in 1911, held a similar line on the importance of practicality over theory. Even though scholars such as "Darwin, Lamarck, Weismann, Mendel, and De Vries have given us much to think about," Buffum reserved his greatest admiration for "men like Nillson [sic], Burbank, [Willet Martin] Hays, and others who have not hesitated to climb over supposed barriers and have given the world increased production of wealth."[31]

Results counted for far more than theoretical coherence among American breeders. On the one hand, this seemed to work in the Mendelians' favor, as their laws and ratios were confirmed and reconfirmed by case studies from across the United States. On the other hand, it created a potentially dangerous situation for the new science of genetics. A single experimental result, or new variety of plant or animal, that did not confirm Mendel's laws could undermine the universal applicability of genetics. By 1909 two Illinois swine breeders, Q. I. Simpson and J. P. Simpson, were so enamored of Mendel's laws that they even labeled them as the biological equivalent of the celestial mechanics uncovered by Galileo Galilei, Johannes Kepler, and Isaac Newton. Now, thanks to the work of Castle on guinea pigs and rabbits, and Spillman on wheat and cattle, Mendel's laws had proved applicable to "every detail in the building of plant and animal." The Simpsons even declared that "we may now say that astronomical law is no more universal than are these laws of genetics."[32]

As exceptions or contradictions to a universal law would be paradoxical, the two Simpsons were keen to follow in Bateson's footsteps and deny the possibility that acquired characters might be inherited. Just as Bateson had declared the old evidence for the inheritance of adaptive changes null and void, so the two swine breeders asserted that the "guesses of Aristotle" were in the process of being "replaced by the [genetic] laws of causation."[33] In fact, the Simpsons were aware of experiments that would lay the neo-Lamarckian argument to rest once and for all. "Dr. C. C. Guthrie . . . is transplanting ovaries from hens of one breed into hens of another and successfully making breed tests by their progeny." This experiment, they stated, "will silence the dogma of acquired inheritance."[34] Unfortunately for the Simpsons, their prediction could not have been more wrong. Guthrie, then a physiologist at Washington University, was transplanting ovaries between different chicken varieties, but his results would not confirm Mendelian laws. Guthrie would instead claim to have created what were essentially graft hybrid animals, launching a debate in which some would see their interpretation of Mendelian genetics at stake.

The Crucial Test of Lamarck and Weismann

"Owing to the uniform results from the controls," argued Guthrie at the end of his 1907 breeding season, "it may be assumed that the strains of chickens used breed true to color. Therefore any variations in the offspring from the operated hens were due to other influences." Guthrie was confident that his ovarian transplants had succeeded and that his chickens were purebred, meaning that any stark changes in their appearance down the generations must have come from these "other influences." He concluded, contrary to Weismann, that the body of the hens with engrafted ovaries had "exerted an influence on the color of the offspring."[35] Given the animated state of debates over the nature of heredity in the United States, however, Guthrie was cautious. He first announced his findings to the American Physiological Society in Washington in May 1907. The young physiologist's results were then reported at the International Congress of Physiologists in Heidelberg in August 1907. When Guthrie finally published an account of his chicken experiments beyond the auspices of the American Physiological Society in 1908, he remained ambivalent on the heredity question: "It seems that the qualities of the ovaries transplanted in these experiments may have been modified by the foster mother. Or the foster mother influence may have only been impressed after fertilization of the egg; or the influence may have been effective both before and after the egg was discharged from the ovary. A discussion of possible ways in which such influence might act would be unprofitable at this time. At present we are justified only in saying that a field appears to be open to attack with a reasonable hope of profit."[36] Guthrie was intrigued by his findings, which seemed to call into question the Weismannian conviction that heredity was confined to the sexual organs. His caution in interpreting the mixed black and white feathers of his chicks was understandable. After all, in the same year that Guthrie had announced his intriguing experimental results before the American Physiological Society, Bateson had used his lecture series at Yale to denounce the implication that heredity might extend beyond the germ cells to the somatic cells. Yet over the course of 1909, Guthrie chose to promote his chicken experiments and their implications for heredity to an ever-growing audience, including the American Breeders' Association. These actions would bring him into contact with William E. Castle, one of the foremost advocates of Mendelian genetics in the United States.

Like Guthrie, Castle began life on his family farm—in his case, near Alexandria, Ohio. His interest in the natural world developed at an early age, when "as a farm boy he collected wild flowers and learned to graft trees and to identify and prepare the bones of animals which had died in fields and woods."[37] Castle entered Harvard with a scholarship in 1892 and worked as a laboratory assistant under Davenport. Following his doctorate, Castle spent only a few years away from Harvard before he was called back to serve at his old institution. He arrived at Harvard's Zoological Laboratory in time for what he deemed "one of the great discoveries in biology," that of the laws of heredity originally revealed by Mendel. Castle believed that the success of Mendel's laws post-1900 was in part due to "Weismann's germ-plasm theory, in particular the idea of the non-inheritance of acquired characters, [which] had put the scientific public into a more receptive frame of mind." Another factor at play was Bateson, who, Castle claimed, had pointed "out the full importance and the wide applicability of the law. This he has done in two recent publications with an enthusiasm which can hardly fail to prove contagious."[38]

Castle's interpretation of Mendelian genetics, from the purity of germ cells to the reality of discontinuous variation, was essentially identical to Bateson's. His interpretation of Weismann's theory, as eliminating the influence of somatic cells or the wider environment on heredity, also imitated Bateson's. So we should not be surprised to find that Castle was unreceptive to Guthrie's transplantation experiments on chickens. Castle was also aware of similar reports from Europe. In late 1907 Professor Vilhelm Magnus, of the University of Christiania (now Oslo), had conducted experiments that were nearly identical to those of Guthrie's, transplanting ovaries between rabbits to obtain mixed-colored offspring. Castle set out to test Guthrie's claims, backed by Harvard's Laboratory of Genetics at the Bussey Institution and a grant from the Carnegie Institution. Alongside his assistant, John C. Phillips, Castle attempted to replicate Guthrie's experiment using guinea pigs. Following advice on transplantation surgery provided by Carrel, Castle and Phillips set to work. They removed the ovaries from an albino guinea pig and transplanted the ovaries of a black guinea pig in their place. The albino guinea pig was then crossed with an albino male and produced two black-colored young. Castle interpreted his results as typical of a Mendelian dominant—in this case, black fur—overcoming an albino Mendelian recessive. He reported in August 1909:

> In all recorded observations upon albino guinea-pigs, of which we have ourselves made many hundred, albinos when mated with each other produce only albino young. Accordingly there seems no room for doubt that in the case described the black-pigmented young derived their color, not from the albino which bore them, but from the month-old black animal which furnished the undeveloped ovaries, for transplantation into the albino. As regards the important question whether, in such an experiment as

this, the germ-cells are modified in character by the changed environment within which they are made to grow, our results are at variance with those of Guthrie and Magnus. *We can detect no modification.* The young are such as might have been produced by the black guinea-pig herself, had she been allowed to grow to maturity and been mated with the albino male used in the experiment.[39]

Castle suspected that Guthrie's chickens might have regenerated their ovarian tissue. The ovaries implanted by Guthrie must have ceased to function, rendering his later breeding experiments and their implications on heredity theory null and void. Castle's guinea pig experiments, on the other hand, by making use of Mendelian dominant and recessive characters, had eliminated such a possibility. Guthrie was not prepared to accept Castle's findings.

Less than two months after Castle's critique appeared in *Science,* Guthrie wrote to the journal on the matter of "guinea pig graft-hybrids." His ostensible aim was to report upon his own attempt to carry out an ovarian transplantation between guinea pigs. Although Guthrie's guinea pig had given birth to two young in August 1909, Guthrie made a show of arguing that his latest transplantation experiments had no bearing on heredity as his animals were not purebred and already had a mixture of different fur colors. Since "all the animals were mongrels [as opposed to purebred animals,] it is obvious that no conclusion regarding foster mother influence is possible." These experiments contributed only to physiology, not genetics.[40] He then argued that it was likewise impossible for Castle to have seen any influence from his guinea pig foster mothers. "Had the operated pig been bred to a male of the same strain as the pig from which the engrafted ovary was obtained," Guthrie pointed out, "then in view of my own results on fowls, and Magnus's results on a rabbit, characteristics in the offspring indicative of such influence might have been obtained."[41]

Guthrie and Castle had now had the first of a series of exchanges on the technical aspects of one another's transplantation experiments. The wider implications of their argument for the future of Mendelian genetics in the United States was becoming clear for all to see. In December 1909 Guthrie addressed the meeting of the American Breeders' Association at Omaha. He began by taking his audience through the nineteenth-century literature on graft hybrids, notably the work of Charles Darwin on plants. From his reading of Darwin, Guthrie concluded that "graft hybrids in plants are by no means rare; and in some species, e.g., the potato, not difficult of realization."[42]

The creation of animal graft hybrids was, of course, more challenging. In 1890 the English biologist Walter Heape had transplanted the fertilized ova of an Angora doe rabbit into a Belgian hare rabbit. Heape conducted his experiment to "determine in the first place what effect, if any, a uterine foster-mother would have upon her foster-children," concluding that as "far as this single case goes, the evidence is negative."[43] Guthrie argued that these earlier experiments, "though brilliant in their conception are open to so many errors," including the possible vagaries of impregnation and contamination, "that we believe any conclusions from them are unwarranted."[44]

Guthrie went on to describe the results of his transplantation work in chickens before the members of the association. With his recent exchange with Castle in mind, Guthrie went out of his way to describe how he had obtained his purebred chickens from a reputable breeder. By this time, Guthrie had dissected his chickens and was able to go into great anatomical detail, demonstrating that—contrary to the conclusion reached by Castle—his ovarian transplants had functioned and that ovarian tissue from his foster hens had not regenerated. On the crucial question of whether his chickens were true graft hybrids, Guthrie demurred. "One should not consider animal offspring from engrafted ovaries as identical with the graft hybrids of plants described by Darwin," he explained, as he had "used the term provisionally and as a matter of convenience." His experiments, however, were heading in the right direction. The next step, Guthrie argued, would be to transplant eggs or embryos between animals. If "such resulting offspring show 'foster' influence, then a certain parallelism will be established between animal and plant 'graft hybrids.'"[45]

Even if Guthrie had carefully avoided labeling his chickens as true graft hybrids, his actions were fast becoming an existential threat to the alliance of the Weismann barrier and Mendelian genetics. In March 1911, Castle and Phillips published a lengthy rebuttal, titled *On Germinal Transplantation in Vertebrates*, under the auspices of the Carnegie Institution. In it, they explained what was at stake in the different interpretations of transplantation experiments:

> The curiosity of zoologists has long been aroused to know whether the reproductive gland of a vertebrate can be successfully transplanted from the body of one individual to another; and, if so, whether the gland will thereafter function in its new environment; and, if it does, whether the nature of its products will remain unaltered. The fact has repeatedly been pointed

out that experiments of this sort, if successful, should afford a crucial test of the Lamarckian and the Weismannian views, respectively, of the relation of the germinal substance to its environment and in particular to the body. ... Since it is known that the environment directly influences the nature of the body, if it can be shown further that the body directly influences the character of the inheritance through the sexual products, the Lamarckian principle is established and that of Weismann is disproved. It is therefore of fundamental importance either to confirm or to disprove the results of the authors mentioned. We are unable to confirm, we present evidence which tends to disprove, the conclusions reached by Guthrie and Magnus. We do not question the results reported by them, but only the interpretations given by them to that work.[46]

Castle and Phillips had essentially depicted Guthrie as a Lamarckian who did not believe in the independence of the germ cells from the wider environment. They also suggested that American breeders and biologists could no longer take a position of neutrality or compromise between Lamarckian and Mendelian interpretations of Weismann's principle. If Guthrie's chickens were to be accepted as graft hybrids, Castle and Phillips argued, this would be the start of a slippery slope that would lead to the complete overthrow of Weismannism. If the somatic cells could influence heredity, then it was also true that the environment could influence heredity through its well-established influence on the body. By taking this stance, Castle and Phillips had taken Guthrie's chickens from a minor, if intriguing, heredity phenomenon to the crucial test of Lamarckism and Weismannism.

Only a few days after the release of Castle and Phillips's report, Guthrie wrote a letter to *Science,* in which he accused his counterparts of stealing his priority in demonstrating that offspring could be produced from transplanted ovaries. He denied being a Lamarckian. "I had no allegiance with any school of theorists," declared Guthrie. "Whether [my] results would substantiate either or neither of the theories built largely upon speculation as to the relationship of reproductive tissues to their environment," he added, "or whether the character of the offspring would conform to Mendel's results of studies of inheritance in peas, gave me no concern."[47] He claimed that he had embarked upon his experiments only to see if engrafted ovaries would retain their function, with later breeding experiments conducted as a matter of pure interest. Guthrie went on to explain that he would have continued his breeding program into the second generation had his ambitions not been "terminated by an outbreak of disease among the fowls." Even had his planned second generation of chicks appeared, Guthrie went on, he would not have considered that they "could do more than indicate whether or not soma influence might be evident." Aware of the limitations of his experiments, he declared, "I saw the limitations to the absoluteness of any evidence that might be obtained by continuation of such experiments."[48] This explains his reticence to draw more than provisional conclusions on matters of heredity.

Castle took his time crafting his response. In July 1911, a month after Guthrie's letter appeared, he offered his rebuttal. He began somewhat disingenuously. "Beyond the point of clearly stating the essential difference in our conclusions and the ground on which this difference rests," Castle began, "I take it [that] neither Professor Guthrie nor I would care to go in the way of discussion." Castle had, of course, cared to go exactly this way in his Carnegie Institution report, when he had lumped Guthrie with the Lamarckians. Castle went on to argue that the crux of their disagreement hinged on whether Guthrie's graft hybrid chickens had come from transplanted ovaries, or from another source:

> The facts are these. [Guthrie] transplanted the ovary of one hen into another hen. The second hen afterward laid eggs. Does it follow that the eggs came from that transplanted ovary? Not unless it can be shown that there was no other possible source from which they could have come. What should we say to this sort of evidence? A boy rushes into the house. "Father," he says, "I have killed a hen."
>
> "How do you know, my son?"
>
> "Why, I threw a stone over the fence into the henyard, and when I opened the gate and went in, there lay a dead hen." Is that proof that the hen was killed by the stone which the boy threw over the fence?[49]

To accept Guthrie's conclusion, noted Castle, we must be completely certain that the ovaries introduced into his hens had survived and that no other ovarian tissue was present in his animals. Castle's experience with guinea pigs at Harvard had convinced him that successful transplants were few and far between. He also doubted that the variation in feather color reported by Guthrie could not be explained through ordinary inheritance. In the view of American Mendelians, it seemed far more likely that Guthrie's hybrid chickens had come from regenerated, not transplanted, ovaries. Castle also countered Guthrie's suggestion that his guinea pigs were not purebred, on the grounds that he had overseen their breeding firsthand. Despite his earlier promise not to engage in such a discussion, Castle addressed heredity theory and dismissed Guthrie's claim to be a neutral party in one of the great questions of early twentieth-century biology:

> But if it be supposed, as Guthrie does, that the ova came from an engrafted ovary, then serious contradictions are encountered as regards the color inheritance. Such contradictions Guthrie may not lightly push aside by disclaiming any interest in laws of inheritance on the ground that they are of "no concern" to him. He who claims to have modified inheritance should know what *normal* inheritance is, and he can not divert attention from chickens by scornful references to "peas," nor from stubborn facts by thrusts at "theories built largely upon speculation." No theories are involved in this discussion except the one which Guthrie has propounded, that inheritance is affected by foster-mother influence.[50]

The final word of 1911 went to Guthrie. Now based at the University of Pittsburgh's physiological laboratory, he chose to end his debate with Castle in much the same way it had begun: on a technical criticism of Castle's transplantation experiments. "Professor Castle has objected to my application of the term mongrel to guinea-pigs used by him in experiments which he claims overthrow my results on chickens," stated Guthrie, even though his "authority for the use of this term is the following extracts from his [Castle's] paper."[51] Guthrie then related how Castle explained discrepancies in the color of his supposedly albino guinea pigs by reference to the spotted colors of the litter from which his albinos had been obtained. At least one of these guinea pigs had not come from a purebred line, concluded Guthrie, implying that Castle's experiments had nothing meaningful to add to the question of somatic influence in heredity. Five years had passed since Guthrie had first recorded some peculiarities in the color of his chickens. As no further experiments took place, the dispute between Mendelian genetics and the graft hybrid had effectively resulted in stalemate.

The Hidden Hand of Charles Davenport

Guthrie's first interaction with Charles Davenport, director of the prestigious Cold Spring Harbor Laboratory and former mentor of Castle, did not bode well for their future relationship. "I have seen an abstract of your paper," Davenport informed Guthrie in July 1907, "[and] I may be permitted to state that the results that you have obtained are exactly such as I would have expected on the hypothesis that the grafted ovaries were completely atrophied." Davenport, like Castle, believed that Guthrie's chickens had rejected their transplants and regenerated their own ovaries. Once this occurred, Davenport argued, there was no reason to be surprised by the appearance of chicks with a mixed plumage—a phenomenon that could easily be explained through Mendelian principles. "I can not therefore accept your conclusion that the grafted germ cells have been influenced by the foster mother," wrote Davenport. "The essential part is to know that the eggs laid came from the transplanted ovaries."[52] It would be easy to think of Davenport as a Mendelian intent on countering Lamarckism, but his letter was only the beginning of an extensive correspondence with Guthrie, in the course of which Davenport would begin to display a conciliatory attitude.

Another product of an American family farm, Davenport had abandoned a career in engineering—favored by his overbearing father—to enter biology. His rise was meteoric. He enrolled at Harvard, submitting his doctoral dissertation in 1892 and devising a new course on experimental morphology in 1893. It was at Harvard that Castle would briefly work under Davenport as a laboratory assistant. In 1898, Davenport was appointed director of the Cold Spring Harbor Biological Laboratory, which would soon become "a hotbed of Mendelian studies of flies, mice, chickens, ducks, canaries, sheep, rabbits, and corn."[53] Davenport did not immediately convert to Mendelism in 1900. His friend and biographer, E. Carleton MacDowell, reported that Davenport had conducted breeding experiments on mice from 1900 to 1904, but failed to spot any patterns in their color that conformed to Mendelian genetics. He subsequently concluded that heredity could not be explained by Mendelian law alone.[54] Despite his skepticism, Davenport was one of the first American biologists to address the rediscovery of Mendel's laws in the *Biological Bulletin* of June 1901.[55]

Davenport's ambiguous take on Mendelian genetics came to the surface in August of 1901, during his vice-presidential address to the American Association for the Advancement of Science. His speech, "Zoology of the Twentieth Century," expressed the hope that biologists would soon solve the mysteries of inheritance and variation, including "the laws of mingling of qualities in hybrids" and "an explanation of the monstrosities and the sterility which accompany hybridization." Davenport thought that cytology would be a necessary part of any explanation for the mysteries of hybridization.

He also hesitated to weigh in on the causes of variation, as "within the last decade a profound student of variation (Bateson) has declined to discuss its causes, holding that we had no certain knowledge of them."[56] Something happened between 1901 and 1907 to change Davenport's opinion of Mendelian genetics. MacDowell declared that a meeting with Bateson in 1907, on the latter's second visit to the United States, was instrumental. Yet Davenport was already an admirer of Bateson and had invited him to Cold Spring Harbor prior to the latter's arrival in New York in July 1907.[57] In all likelihood, Davenport did not receive some sudden revelation, but gradually swung to the Mendelian viewpoint sometime after 1904. An important moment may have been the presence of Hugo de Vries at the opening of Cold Spring Harbor in 1904. De Vries emphasized the importance of studying mutations and their ability to create something akin to Bateson's "discontinuous" inheritance: allowing organisms to jump from one species to another in a single generation.[58]

Despite expressing his official disapproval at Guthrie's intervention into heredity theory, Davenport soon softened his tone. In January 1908, six months after the dispatch of his first letter, he wrote to Guthrie again, this time inviting him to Cold Spring Harbor that summer. "I could afford you material for grafting experiments," wrote Davenport, "and there are certain other experiments that I should like to cooperate with you in doing, likewise involving surgical operations on the abdomen."[59] Guthrie quickly responded, thanking Davenport for his kind letter and making arrangements for his absence from the Washington University Medical School. He also expressed great excitement at the prospect of working with Davenport at Cold Spring Harbor, by now the site of the Carnegie-funded Station for Experimental Evolution. "I am greatly pleased with the opportunity of working under your direction and trust my services may prove of some value to you," wrote Guthrie. "It will be possible to perform the operations you desire on ovarian transplantations by the middle of May. I shall be very glad to cooperate with you in any other experiments that you may have planned."[60] Davenport elaborated further on these planned experiments in a letter of March 6, 1908: "I neglected to mention in my last letter that the most important thing to do now is to perform some experiments of injection of some salts into the ova of laying hens to see if such injection would have an effect on their laying capacity. My hens are laying now very rapidly and we have plenty of material for this series of experiments and by the time you come again in May, we will have plenty of pullets for the transplantation work."[61]

Guthrie's visit to Cold Spring Harbor was much shorter than planned, thanks to an administrative shakeup at Washington University following the death of its dean. Guthrie visited Davenport in early April and carried out salt injections on the Station for Experimental Evolution's chickens. The short visit meant that their planned transplantation experiments had to be abandoned. Guthrie, however, was pleased with the experience. On April 15, 1908, he wrote to Davenport to thank him for his hospitality and to describe the results of his own chicken experiments. "I have just opened some incubated eggs from a black hen," he explained, "carrying a white ovary and bred to a black rooster. Two of the eggs contained dead embryos *almost* sufficiently developed to show the feather markings. More eggs are being saved for putting under a hen."[62] Davenport was also pleased. That May he once again contacted Guthrie, this time to propose a permanent collaboration: "I would suggest the following proposition for your consideration. You to work at Cold Spring Harbor in connection with this Department during such part of the year as you can. The work to be on the physiology of reproduction, using the method of transplantation and to cooperate with me on certain experiments in modifying the chemical composition of the eggs in the ovary."[63] Davenport had a rationale for building a closer working relationship with Guthrie. On a personal level, Davenport had clearly been impressed when he met Guthrie at Cold Spring Harbor in April 1908. Equipped with only the most basic surgical equipment, Guthrie had swept into the Station for Experimental Evolution and performed Davenport's experiments quickly and efficiently. There was perhaps another factor in Davenport's mind. Perhaps Davenport hoped to steer Guthrie's interest and technical expertise into less controversial areas of research. The "physiology of reproduction," working out how fertilization and embryology functioned, could yield practical benefits without any troublesome implications for the nature of heredity.

Sadly, the collaboration was not to be. Guthrie felt that the offered salary was insufficient and requested an additional two hundred dollars to employ an operative assistant.[64] Davenport was unwilling or unable to raise the funds required to draw Guthrie away from Washington. Still, their relationship was largely unaffected. "Although our plans did not work out," remarked Davenport in a letter of May 1908, "you may be sure I shall have the greatest interest in your work and that you will let me have your reports as soon as issued or direct me to the place where they are published." He also reported that the chickens he and Guthrie had operated on with salt injections in April had not produced any eggs.[65] When Guthrie replied in late August, it was to report a series of disasters. His wife and child had been taken ill and the new dean of the Washington University Medical School had slashed the number of students under Guthrie's charge, leaving his laboratory undermanned. He had also moved his precious graft hybrid chickens to his family farm, where they "died of a rapidly fatal infectious disease." This mishap could not have come at a worse time. His first papers on ovarian transplantation had just appeared, alongside a rebuttal from Castle. Their dispute had caught the attention of Spillman, the member of the US Department of Agriculture who had argued that Mendel's laws could be applied to cattle breeding. Spillman planned to meet Guthrie in the hopes "that it would be possible to arrange experiments that will settle the matter definitively."[66]

With the loss of Guthrie's chickens, the experimental showdown envisioned by Spillman never materialized. Davenport was sympathetic to Guthrie's plight, but not supportive of his ideas. "I was glad to receive your letter of August 22nd," he wrote in response, "and I am very sorry to learn that you lost so many chickens after removing them to the farm." Davenport had read Guthrie's 1908 paper in the *Journal of Experimental Zoology*, but was not convinced that Guthrie's late chickens had been graft hybrids: "I still feel that your results are capable of explanation on grounds of regeneration of the normal ovaries."[67] Guthrie did not give in. He perhaps sensed that this was an important moment. If he could not succeed in convincing a sympathetic and collegial Mendelian like Davenport to take his experiments seriously, his cause was as good as lost. He spent almost a year crafting his response. "I was prepared for re-declaration of your non-acceptance of the results of the ovarian grafts," Guthrie responded on April 14, 1909. "Since I do not believe you would ever be convinced by the results obtained by breeding, I will present the following new facts." One fact that caught Davenport's eye read as follows:

> 3. In one hen that laid with a transplanted ovary, that at post mortem in addition to the large functioning ovary, which was well forward as regards the normal site, a small mass of ovarian tissue *made up of small follicles only* was found situated posteriorly to the chief mass, and entirely separate from it. This is, I think, in confirmation of the view that I have expressed that *should any ovarian tissue escape removal*, it would be of the comparatively undeveloped type, and the probability would be that it would not mature so soon as the transplanted *riper* tissue.[68]

Guthrie reasoned that the original ovaries he had removed from his chickens were unlikely to regenerate. Even if they did so, their recovery would be so slow that it would have been impossible for them to regain their functionality during the course of his transplantation experiments. He therefore saw the charge of Davenport and Castle—that his patterned chicks had not come from transplanted ovaries—as groundless. Confronted by the new evidence from Guthrie's dissected hens, Davenport wavered. He asked Guthrie if he could be certain that the large functioning ovary was the transplanted organ. "Is it not probable," he inquired, "that the transplantation of an ovary provides stimulus to the regeneration of the normal ova?"[69] He also suggested that Guthrie conduct further grafting experiments by transplanting the ovaries of White Leghorn chickens into Dark Brahma hens. Davenport knew that the white plumage of the Leghorn was a dominant Mendelian trait. If the hens were crossed with Dark Brahma males and all the resulting chicks were white, this would once and for all demonstrate that transplanted ovaries functioned while the original ones did not regenerate.

A little over a week later, Guthrie responded. He dismissed the possibility that transplantation might stimulate the generation of other organs, pointing out that degradation was the more likely outcome. He also reported that he had carried out several more grafting experiments in different chicken varieties, but was not hopeful of obtaining eggs. Finally, for the first time, Guthrie acknowledged his growing fascination with the concept of graft hybridization in botany: "In closing I shall mention some similar work that I have puzzled over for a long time—graft-hybrids of potatoes. This work seems to have been done so frequently, and the results have been so consistent, that it would seem that there can be no doubt of the influence of the *tuber* over the engrafted bud's offspring. As I am not a botanist, I have refrained from drawing a definite conclusion—or of using the results in connection with the results on fowls."[70] Davenport did not mention Guthrie's growing interest in graft hybrids in their personal correspondence. Much of their communication over the remainder of 1909 was instead devoted to obtaining Carnegie funding for Guthrie's research. Guthrie sent manuscripts to Davenport, including his paper on graft hybrids delivered to the American Breeders' Association. Davenport even edited Guthrie's first response to his Mendelian critics.[71] By the end of 1909 the fast and sometimes furious correspondence between Guthrie and Davenport had begun to peter out. Both men would remain in touch for years to come, sharing poultry catalogues and exchanging reprints of scientific papers. Both also continued their transplantation work, but never saw eye to eye.

When Davenport publicly addressed Guthrie's experiments in January 1911, he maintained that their results could best be explained by the regeneration of the hen's original ovaries. Davenport's own efforts to replicate Guthrie's hybrids at Cold Spring Harbor had met with repeated failure. "With the [Cold Spring Harbor] results," declared Davenport, "the data of Dr. Guthrie's paper are not in disaccord. His data, like ours, furnish no evidence for the survival of the engrafted ovaries, far less of an effect of the soma of the foster-mother on the introduced germ plasm."[72] Guthrie did not hold a grudge. In a letter of June 20, 1911, he thanked Davenport for sending him some reprints. "I am, of course, particularly interested in your work on ovarian transplantation," he wrote, "and though your views and mine do not coincide on certain interpretations, yet I am glad you have undertaken the work at the Station."[73] There were no references to the repeated clashes between Guthrie and Castle. Davenport dispatched a copy of Castle's *On Germinal Transplantation in Vertebrates* to Guthrie in November 1911, but made no further attempts to dissuade the young physiologist from his unorthodox views on heredity.

Davenport remained silent during the very public exchanges between Guthrie and Castle. He perhaps felt that his printed rebuttal of Guthrie's graft hybrids had been a sufficient intervention. Davenport may also have yearned for future collaborations, still hoping to claim Guthrie's esteemed surgical skills for Cold Spring Harbor. Nor was he always supportive of Castle's work. In his 1909 book on inheritance in domestic fowl, Davenport was highly critical of the work of Castle on hybrid crosses in chickens. "I think Castle's paper may justly be criticized," remarked Davenport, "for not giving sufficient data concerning the ancestry of the individual mothers used. Without such data the paper can not be said satisfactorily to demonstrate his conclusion."[74] Intriguingly, Guthrie would also criticize Castle for not keeping track of the ancestry of his laboratory animals. Perhaps Davenport saw the experimental accuracy offered by Guthrie as a greater virtue than the theoretical orthodoxy embodied by Castle.

Guthrie was highly cautious of drawing strong inferences from his chicken experiments. It is equally possible that Davenport remained hopeful that he would change his mind. In the published version of his remarks to the 1909 meeting of the American Breeders' Association, Guthrie began to walk back some of his more ambitious arguments, including the analogy between

FIGURE 3. Variegated oranges. For American agriculturalists and horticulturalists, results mattered more than theories. Selective breeding, hybridization, and exposure to different climates were all believed to produce heritable changes in plants. Grafting was also believed to induce dramatic effects, as demonstrated by fruits with segments of different skins. J. Eliot Coit, *Citrus Fruits: An Account of the Citrus Fruit Industry with Special Reference to California Requirements and Practices and Similar Conditions* (New York: Macmillan, 1915), 122.

his chickens and plant graft hybrids (for a contemporary example of the latter, see fig. 3). "One should not consider animal offspring from engrafted ovaries," he wrote, "as identical with the graft hybrids of plants described by Darwin." Guthrie concluded on the tantalizing remark that, from a "physiological standpoint," nutrition was a factor of "prime importance" to consider in future research.[75] Guthrie had probably been convinced—or had convinced himself—that the mix of black and white feathers on the offspring of his hens could also be explained by their exposure to certain environmental factors during their upbringing. Diet can induce noticeable changes in the appearance of chickens. In a 1949 attack on the inheritance of acquired characters, British biologist Julian Huxley noted that, if fed a diet of corn, the feet of some fowl take on a yellow tinge. Such environmentally induced changes, he argued, are not heritable and are entirely compatible with Mendelian principles.[76]

Experimenter's Regress

By the end of 1911 Guthrie had abandoned his efforts to convince the biological community that heredity could not be explained through Mendelian genetics alone. He instead returned to his far less controversial and higher-status work in medicine. From an outsider's perspective, Guthrie's graft hybrid chickens never presented any genuine threat to the Mendelians. Guthrie conducted only a small experiment, which was cut short after a single generation thanks to the loss of his chickens to disease. Practitioners of Mendelian genetics, on the other hand, reveled in detailed bookkeeping, statistics, and large-scale breeding across multiple generations. Nevertheless, there were some similarities between the ideals espoused by the Mendelians and Guthrie's graft hybrid experiments.

Although Guthrie was unable to conduct large breeding programs, he recorded his methods and results in great detail. Guthrie was certainly better at this than Castle, who found himself criticized by Davenport, a scientist ostensibly on his own side, on the grounds that Castle kept shoddy records. Guthrie's meticulous records, experiments, and use of chickens as a model organism essentially aped the most cherished methods of Mendelian genetics.[77] When he threw down the gauntlet by announcing that the body of the animal could exert an influence on heredity, his challenge could not therefore be easily dismissed as pseudoscience, amateurism, or quackery. If we bear in mind Guthrie's surgical expertise, it comes as no surprise that Castle focused his criticisms not purely on experimental techniques but rather on his perceptions of Guthrie's ideological stance and logical fallacies.

Why was it that some Mendelians, including Bateson and Castle, were so alarmed by the slightest hint of the inheritance of acquired characters? During the early years of the twentieth century, heredity—and perhaps more importantly, its application to agriculture—was an open contest. Mendelian genetics had only a tenuous hold of the hearts and minds of American farmers and breeders. As we have seen, the members of these professions held little respect for theoretical schools of thought. Practical results on the farm mattered above all else. It was largely irrelevant whether improved varieties of plants and animals were obtained through Mendelian crossing, mutation, changes to the environment, or grafting. Although farmers and breeders held a certain flexibility toward heredity theory, many early Mendelians did not. Key players, such as Bateson and Castle, saw the Weismann doctrine as indispensable to the success of their science. Lamarckian ideas and graft hybrids could undermine Weismannian theory and bring down their specific interpretation of Mendelian genetics. Their subsection of the new discipline was vulnerable, with valuable positions and resources to protect.

This interpretation is lent further weight by the letters exchanged by Guthrie and Davenport. We might expect Davenport, as one of the earliest advocates of Mendelian genetics in the United States, to be skeptical of Guthrie's views on heredity and graft hybridization. Although Davenport could never officially agree with Guthrie, in their private correspondence he was quite open to new ideas. From 1908 to 1909 Davenport repeatedly attempted to work with Guthrie. He offered paid visits and summer work at Cold Spring Harbor and sought to direct Guthrie's surgical expertise toward physiology. Unfortunately for Davenport, he was unable to tempt Guthrie to stay at Cold Spring Harbor Biological Laboratory. For his part, Guthrie was able to provide compelling evidence that his ovarian transplants had worked. Davenport ultimately took the line that the removed ovaries of chickens could successfully regenerate and supplant transplanted organs. He did not challenge Guthrie after the publication of Guthrie's 1911 article, with the two men maintaining a friendly correspondence for years after the hybrid chicken controversy.

The fact that a consensus was never reached over Guthrie's chickens can be explained with reference to "experimenter's regress." This regress typically occurs following an empirical experiment aimed at detecting "a novel phenomenon," such as the blurring of the Weismann barrier following ovarian transplantation in chickens or guinea pigs. When it is not clear whether "detection or non-detection of the phenomenon" proves the experiment a success or failure, arguments can occur over "what comes to count as a 'well-done experiment.'"[78] In the case of Guthrie and his interlocuters, a great deal of ink was subsequently spilled over the technicalities and design of transplantation experiments. Questions were raised surrounding breeding and surgical expertise, while Castle even examined the logic of cause and effect on which Guthrie's conclusions were based. Ultimately, a form of scientific "social negotiation" is required to halt the regress by establishing what kind of experiments are valid.[79] This negotiation did not occur in early twentieth-century biology. A shifting scientific landscape and the loss of Guthrie's chickens meant that participants in the debate turned to other matters. Guthrie's claim that the body of the organism had some influence on heredity left little lasting influence.

As the validity of transplantation for informing heredity thinking was never settled, such experiments recurred across the twentieth century. In 1925 Francis Albert Eley Crew, director of the Animal Breeding Research Department at the University of Edinburgh, drew upon experiments with the ovaries of guinea pigs for the third volume of the *Biological Monographs and Manuals* series on animal genetics. If the ovaries of a white-furred guinea pig were transplanted into a red-furred guinea pig prior to crossing with a white male, the resulting offspring would be largely white, albeit with a few red hairs here and there. Crew used this data as part of a wider argument allowing for the inheritance of acquired characters, although he believed that graft hybridization could at best produce heritable changes of a "very restricted character."[80] Guthrie's experiments would reemerge during the same period, in the writings of neo-Lamarckian Paul Kammerer. However, Guthrie's temporary hiatus from the scientific world came about in the wake of a German botanical controversy that threatened the very existence of graft hybrids.

2

Rise of the Chimera

⟨decorative flourish⟩

JUST AS HIS CORRESPONDENCE WITH CHARLES CLAUDE GUTHRIE BEGAN
to peter out, Charles Davenport had another brush with a graft hybrid. In
November 1910 Davenport wrote to George Harrison Shull, an American
botanist and geneticist. The subject of their discussion was a set of photographs of a tomato graft hybrid. "I am sending you the prints that were sent
to me," wrote Davenport, "as I think it is a rather interesting experiment."[1]
Shull, however, claimed to be unable to see any difference between ordinary
tomatoes and those of the plant in the photographs. He also interrogated
Davenport on the life history of the plant, asking how it came to be hybridized and what evidence there was that graft hybridization had taken place.
Shull summarized his skeptical letter with a series of references to authorita-
tive German botanists. "The recent work in graft hybrids by Winkler, Baur
and Strasburger," he concluded, "have been of extreme interest and have
placed Graft Hybridization on a new and sounder footing than it has had
until Winkler's work appeared about two years ago."[2] With that, Shull and
Davenport's discussion of graft hybrids came to a sudden end.

Shull had referenced a body of German scholarship in his brief exchange
with Davenport. Although German interest in Gregor Mendel's work blos-
somed rapidly after 1900, an air of mystery still surrounded Mendelian
factors, termed *genes* by Danish botanist Wilhelm Johannsen in 1909.[3] Into

this arena of uncertainty stepped Hans Winkler, professor of botany at the University of Tübingen, who grandly announced his intention to investigate heredity through experimentation. Winkler began his campaign by grafting black nightshade (*Solanum nigrum*) and tomato plants (*Solanum lycopersicum*) together. Over two hundred crosses later, all made under rigorous experimental conditions, Winkler was left with several grafted plants of particular interest. On five of his grafted plants, a new shoot erupted from where they were joined. The leaves on these shoots were divided into two sections. One half resembled the leaf of a tomato plant and the other nightshade. Winkler described some of these creations as "chimeras," a phenomenon where two intertwined, but genetically distinct, organisms share the same body. He also claimed, however, to have created genuine graft hybrids.[4] His readers, who included botanists from Britain and the United States, were therefore left with a conundrum. Had Winkler actually managed to make genuine graft hybrids, or were all of his grafted plants chimeras? The battle lines for a scientific controversy had now been drawn. Depending on allegiance, one could interpret Winkler's plants one of two ways:

1. *Chimeras.* Generally favored by the Mendelians, including William Bateson, this stance held that all of Winkler's plants were chimeras. Chimeras consisted of two or more conjoined organisms. This combination, although striking to the naked eye, was superficial. The cells of the different organisms present in a chimera did not merge, hybridize, or exchange genes. As a result, chimeras were not true hybrids and did not violate the Weismann barrier. They had no real implications for the science of heredity.

2. *Graft hybrids.* Winkler and his supporters, meanwhile, believed that he had also combined the bodies of different plant species to create true graft hybrids. These plants had exchanged genes, most likely through cell fusion at the graft junction, resulting in plants that were indistinguishable from those created by sexual hybridization. The Weismann barrier had been breached. Graft hybridizers pointed to the unexpectedly high chromosome count of Winkler's most famous plant, the *Solanum darwinianum*, as evidence that hybridization had occurred at the cellular level. However, exactly how many chromosomes the *Solanum darwinianum* had, or how many were required to assure its hybrid status, was a source of ongoing confusion and contention for those involved in the controversy.

One Mendelian who seized on the chimera concept as a means of explaining the strange appearance of some grafted plants (without violating the

Weismann barrier) was German botanist Erwin Baur. Baur had originally studied psychiatry at the University of Kiel but shifted to botany following a series of private lessons and a stint at Kiel's marine zoological institute. In 1903 he was granted an assistant position at Berlin University's Botanical Institute, where he became a stout defender of Mendelian genetics. Winkler's chimeras caught Baur's attention. Among the latter's newfound botanical interests were geraniums with white-edged leaves. Baur observed that the white sections of these leaves were made up of colorless cells, while the green sections were made of green cells. Using a microscope, Baur noticed that these cells did not intermingle, but rested side by side. He also noticed that colorless cells in geraniums were descended only from other colorless cells, while green cells emerged only from other green cells. This suggested that the different cells were genetically distinct.[5] All in all, Baur thought, his geraniums were very similar to the plants created by Winkler. If the two were analogous, then plants that appeared to be hybrids on the surface were not always so at the genetic level.

Baur would eventually win more and more biologists to his way of thinking, with historical examples of graft hybrid plants being "outed" as chimeras. Yet the path to Baur's triumph was long and convoluted, as Winkler continued to argue that he had created true graft hybrids alongside his chimeras. Winkler initially attracted supporters, spurred by the news that one of his plants had a high chromosome count, indicating that hybridization had taken place. Meanwhile, Baur entered a scientific partnership with Bateson. From his new home at the John Innes Horticultural Institution in England, Bateson sought to understand the nature of the chimera and use the concept to explain heredity phenomena in plants. Scientific debate over the graft hybrid ground to a halt with the outbreak of the First World War. In a disaster uncannily like the fate of Guthrie's chickens, Winkler lost his prized example of a graft hybrid, the *Solanum darwinianum*, during the chaos of war. As a result, for a growing number of geneticists, botanists and plant physiologists, the chimera would effectively supplant the graft hybrid as the favored explanation for the intermingled appearance of some grafted plants.

An American Lamarckian in Germany

Graft hybrids occupied a contentious position in American genetics and breeding during the early years of the twentieth century. Even the most stalwart advocates of Mendelian genetics could be seduced by the lure of graft hybrids. In 1906, for example, a guide to grafting in a horticultural encyclopedia contained a conflicting mess of facts and opinions around the possibility of hybridizing plants through grafting. Its author was none other than Liberty Hyde Bailey, professor of horticulture at Cornell University, who had prepared the way for Bateson's rapturous reception at the New York Horticultural Congress in 1902.[6] Bailey wrote that the experiments of Lucien Louis Daniel, professor of agricultural botany at the Lycée de Rennes in France, had shown "that the stock may have a specific influence on the [s]cion, and that the resulting characters may be hereditary in seedlings." He did not, however, expect the possibility of graft hybridization to "modify the practices of horticulturists nor greatly change our ideas respecting the results to be obtained from accustomed operations."[7] Bailey thought that graft hybrids were real, but of limited utility. If one of the earliest advocates of Mendelian genetics in the United States could display such flexibility, it is little wonder that genetics was not the only tool readily available to breeders in American agriculture.

Nor was Daniel the only example of European influence on American botany when it came to graft hybrids. Shortly after Bailey's encyclopedia was published, Winkler first went public with his own graft hybrids, the existence of which he announced in a series of articles in the prestigious *Berichte der Deutschen Botanischen Gesellschaft* from 1907 to 1908. News soon traveled across the Atlantic, intriguing American biologists. One of the most excited was American botanist Douglas Houghton Campbell. Campbell already had close links with the centers of German biology. While studying for his doctorate at the University of Michigan, which he received in 1886, Campbell lived at his family home and saved his salary, all for the goal of embarking on a trip to Europe. Once there, he studied under the famous botanist Eduard Strasburger at the latter's laboratory in Bonn, Germany. Campbell saw how German zoologists sliced their biological samples into thin sections before embedding them in paraffin for examination under the microscope. Impressed by this new way of seeing the inner workings of organisms, Campbell introduced the technique to the United States and applied it to botany.[8] By 1891, Campbell was head of the Department of Botany at Stanford University, a post he still held when news of Winkler's graft hybrids reached him.

Strasburger, Campbell's old mentor, had already come out against the existence of graft hybrids. During a cytological study of a nineteenth-century graft hybrid, the *Cytisus adami* (fig. 4), Strasburger had counted the chromosomes in its cells and noted that the same number occurred in both parent plants. If the plant had been a true hybrid, Strasburger reasoned,

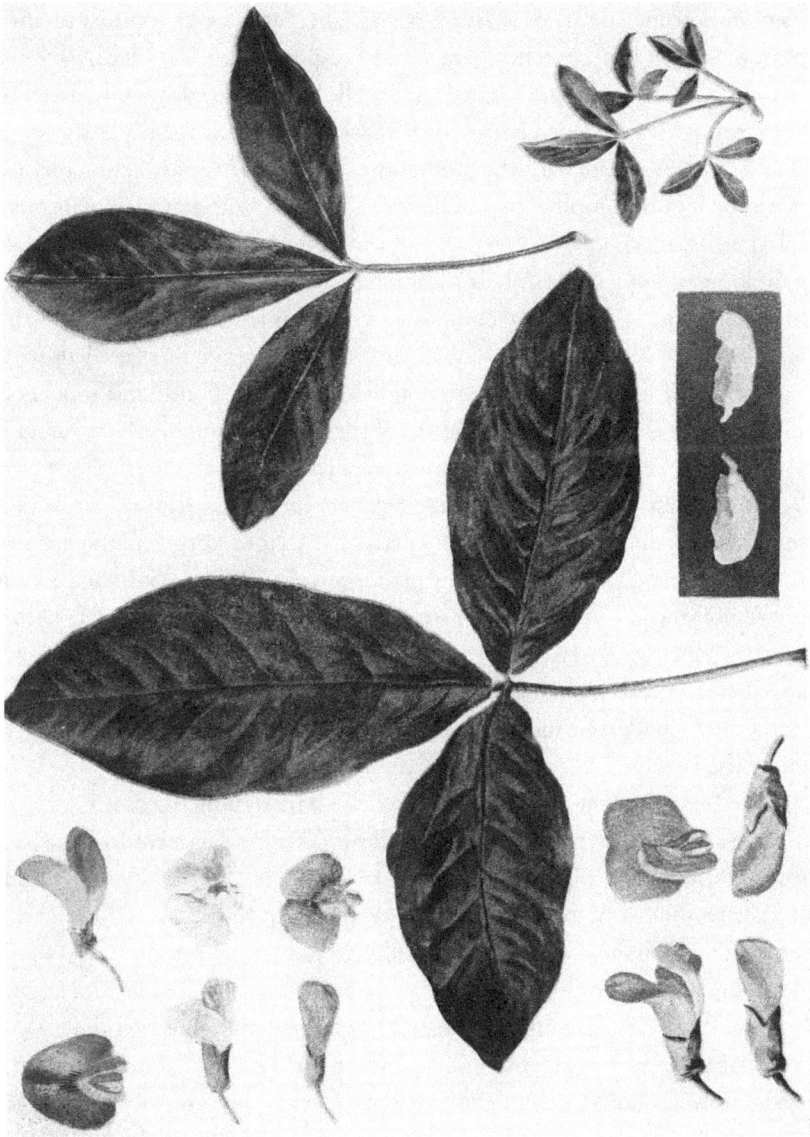

FIGURE 4. *Cytisus adami*. The *Cytisus adami*—created by grafting common and purple laburnum together—had provoked scientific curiosity since its appearance in a plant nursery in nineteenth-century France. In the early twentieth century, the Polish-German botanist Eduard Strasburger was able to count its chromosomes and show that it contained the same number as its parent plants. This indicated that *Cytisus adami* was a chimera and not a true hybrid. Lucien Louis Daniel, *Les mystères de l'hérédité symbiotique: Points névralgiques scientifiques, pensées, théories et faits biologiques* (Rennes: Roger Gobled, 1940), plate X.

its chromosome count would have been higher, as it would consist of chromosomes from both parents. Strasburger responded to Winkler by arguing that his graft hybrids showed no sign of cellular fusion. Ultimately, a kind of compromise was reached between the two botanists. Strasburger theorized that Winkler's plants with the most compelling intermediate characteristics might be highly complex "hyperchimeras," which existed in a "very intimate relationship, such as between parasite and host." Winkler also chose to take a diplomatic approach to their disagreement. Although he did not disown his graft hybrids, he did conclude that hyperchimeras might exist and that Strasburger had been right to label the *Cytisus adami* a chimera all along.[9]

Campbell, however, was not content to rely on secondhand reports or suppositions, no matter the authority of their source. During the summer of 1910 he returned to Germany to investigate the facts for himself. Since 1907, Winkler's first graft hybrid, named *Solanum tubingense* after his university town, had been propagated, with cuttings distributed to botanic gardens across Germany. Campbell found a plant growing from one of these cuttings in the botanic gardens at the University of Munich. Viewing a graft hybrid for the first time, he described the plant as "intermediate in external appearance between the nightshade and tomato," with flowers "larger than those of the nightshade but much smaller than those of the tomato, but like the latter the flowers are a pronounced yellow."[10]

The Munich plant certainly displayed some intriguing intermediate characteristics, but its hybrid status was in doubt. Campbell concluded his visual investigation of the plant by examining its fruit. He described it as dark and similar to that of nightshade, but larger and with flashes of red indicating its tomato heritage. Winkler, reported Campbell, had attempted to rear a second generation of his graft hybrid from the seeds of *Solanum tubingense*, but his efforts had been in vain. Winkler explained his failure to produce and rear viable seed by appealing to his plant's hybrid nature. Tomato fruit and seeds, Winkler noted, take much longer to develop than those of nightshade. Since the fruits of his graft hybrid most closely resembled nightshade, Winkler concluded that he might have planted the seeds of *Solanum tubingense* too early, in the assumption that they had developed at the same pace as those of nightshade, rather than tomato.[11]

Campbell felt unconvinced. His firsthand examination of the Munich hybrid had shown nothing that would distinguish a true hybrid from a chimera. He therefore felt compelled to side with Baur and label the plant a periclinal chimera: "Thus *S. tubingense* has its epidermal region derived from the tomato while the inner tissues including those which give rise to the sporogenous [spore-producing] cells are of nightshade origin."[12] Winkler had not put all his eggs in one basket, however. By the time Campbell arrived at the Munich botanic gardens, Winkler had succeeded in producing several more examples of graft hybrids. His *Solanum darwinianum*, another cross between tomato and nightshade, appeared particularly promising. When Winkler examined the prepared slides of the plant's cells under a microscope, he counted twenty-four chromosomes. He declared that these chromosomes were originally derived from germ cells with forty-eight chromosomes, with half lost following some kind of cell division or "shedding." The *Solanum darwinianum* had therefore received its forty-eight chromosomes from its nightshade parent (contributing thirty-six chromosomes) and tomato parent (contributing twelve). Not everyone was won over by this tortuous explanation, but an excitable Campbell was. He subsequently parroted Winkler's chromosome argument and translated his new classificatory scheme for hybrid plants for an American audience. The star of the show, *Solanum darwinianum*, was placed in a group cumbrously titled "Fusion graft-hybrids arising from a fusion of two somatic cells derived from distinct species."[13]

When Campbell returned to the United States and described his botanical travels in the *American Naturalist* in January 1911, he declared, "These remarkable experiments of Winkler's must be of the greatest interest to all students of the problems of heredity." In retrospect, he mused, it would come as no great surprise to his fellow botanists that plants were able to exchange hereditary information through the fusion of their somatic cells. "It must be remembered," noted Campbell, "that in the evolution of the higher plants there has been a constant tendency toward a reduction of the sexual reproductive parts." So great was this tendency, in fact, that "Many biologists quite ignore the fact that the flowering plant, as it is generally understood, is a purely sexless organism."[14] So why had Campbell been so completely won over to Winkler's cause, especially when his encounter with a graft hybrid in the Munich botanic gardens had proved a disappointment? Perhaps his formative years in science, working with microscope slides in Bonn, inclined him to believe that a careful chromosome count carried out by a studious German professor could not possibly be wrong.

Campbell also had his own ideological perspective on botany and heredity. He praised the theories of Jean-Baptiste Lamarck and believed that the characteristics acquired by organisms during their lifetime could be passed on through the generations. In October 1911, several months after he had described his travels in Germany and Winkler's graft hybrids, Campbell published a volume titled *Plant Life and Evolution* for the American Nature Series. In it, he condemned neo-Darwinians who denied the inheritance of acquired characters. Campbell instead favored Lamarck, "who has been valiantly defended by a host of ardent advocates during the past generation."[15] He noted that European fruit trees that arrived in the United States produced new forms when exposed to different climates, while American growers often sent their flower and vegetable seeds to California to accelerate the growth rate of later generations.[16] Although he claimed to see the value of Mendel's work, Campbell proclaimed "numerous exceptions" to the laws of Mendelian genetics, "for which as yet there is not an adequate explanation."[17] For a neo-Lamarckian like Campbell, graft hybrids provided living proof that characters could be exchanged between species during their lifetime and passed down onto future generations.

Another early enthusiast for Winkler's graft hybrids was the Canadian botanist Reginald Ruggles Gates. Born in Nova Scotia in 1882, Gates grew up on a large family farm full of medicinal plants for his father's apothecary. He graduated in the sciences from Mount Allison University in 1903 before heading to McGill University, peering through microscopes and collecting fossils and fungi as he went. A fellowship at the University of Chicago gave him the opportunity to count the chromosomes of Hugo de Vries's *Oenothera* and interpret the confusing heredity of the plant through a Mendelian framework. By 1910, Gates had embarked on a European tour, which included a stop in Amsterdam, where he met de Vries, before settling in London.[18] From his base in England, Gates took it upon himself to keep readers across the English-speaking world informed of the graft hybrid debates in Germany. Writing in the *Botanical Gazette* in January 1909, Gates introduced his readers to Winkler and how he had produced graft hybrids:

> The method is to graft one species on another in the ordinary manner, and after the scion has "taken," to sever the stem at a point where the tissues of both species will be cut. Adventive shoots then grow out from this cut surface. These will have the characters of either species according to the point they grow from. Shoots arising from the point of contact of the two species gave a peculiar result, which may be described. A scion of *Solanum nigrum* was grafted in this way on a seedling of *S. lycopersicum*, and the shoot in question, originating from the point of contact of the parental tissues, bore

leaves having on one side of the stem the characters of S. *nigrum*, and on the other side those of S. *lycopersicum*.[19]

Gates hedged his bets on whether such plants were true hybrids, noting that Winkler had described them as chimeras. Winkler "concludes that the cells of two different species may come together in other than a sexual way," Gates explained, "and thus serve as the starting-point for an organism which shows simultaneously the characters of both parent species."[20] Whether these organisms combined the genes of their parents or kept them separate he left as an open question.

Two months later, in March 1909, Gates joined the Winkler camp. He recognized that Winkler's earlier experiments, the results of which had been published in 1907, had produced chimeras. Winkler had, however, carried out a fresh set of grafting experiments, with astonishing results. "He has finally succeeded," wrote Gates, "in producing a true graft hybrid." Winkler had grafted two hundred and sixty-eight specimens of S. *nigrum* and S. *lycopersicum*, creating around three thousand promising-looking shoots. Of these shoots, Winkler identified five as chimeras and one as a graft hybrid. Importantly, noted Gates, he had used pure lines to obtain his parent plants, imitating the methods used by Mendelians to obtain their own crosses. "Several interesting cytological questions," mused Gates, "which Winkler hopes to determine, are involved in the nuclear and chromosome behavior of his graft hybrid." As for how the graft hybrid had been created, Gates theorized that "there must be a union of cells, nuclei, or chromosomes, or perhaps of all three, in the production of this form."[21]

Gates was not only occupied by the latest twists and turns in the Winkler story. He also found time to review Bateson's latest book on Mendelian genetics, *Mendel's Principles of Heredity*. Gates was largely positive about the book, noting that it would be of use to "practical breeders of plants and animals." He was less impressed, however, by Bateson's exclusion of any form of heredity beyond Mendelian genetics: "It is a curious blindness to other facts of heredity which leads the author to the opinion that Mendelism probably represents the only type of inheritance which exists. Because characters sometimes behave as units does not exclude the occurrence of several other types of hereditary behavior, nor does the recognition of this fact belittle the facts of Mendelism."[22] Gates clearly took a different line from those Mendelians who saw adherence to the Weismann doctrine as essential. Alternative mechanisms of heredity, thought Gates, were not incompatible with Mendelian genetics. Perhaps the intriguing prospects opened by Winkler's creation of a graft hybrid crept into the young biologist's mind as he wrote his review. At the end of 1909, Gates returned to the graft hybrid controversy. He recounted that Winkler had produced several more graft hybrids, including the *Solanum darwinianum* that had won over Campbell. The plant, recorded Gates, had been obtained by repeatedly decapitating a branch from one of Winkler's chimeras, until only "a pure shoot remained."[23] The hybrid character of the plants, Gates argued, was confirmed by their intermediate characteristics. They could not, therefore, be dismissed as the product of some unique mutation.

For almost a year, Gates had faithfully translated and articulated Winkler's findings. By the summer of 1910, however, the German botanist was beset by critics. The second generation of his plants reverted to their parent types, robbing him of compelling proof of hybridization. Gates subsequently found his faith wavering, not helped by the fact that he misinterpreted Winkler's chromosome count as proof against the hybrid status of *Solanum darwinianum*. Gates also reported that Strasburger had attempted to carry out his own grafts with the aim of searching for instances of cell fusion. Strasburger found nothing of note, "but of course," Gates argued, "negative evidence in such a matter is inconclusive, for the graft hybrids are rare at best." Baur was also on the attack, having "reiterated recently his belief that these forms are explainable as periclinal chimeras, varying in the arrangement of the layers in the growing point."[24] Gates indicated that the array of chromosome numbers produced during the various counts of Winkler's graft hybrids added to interest in the ongoing controversy. These counts, however, only produced confusion as to exactly which plant had which number of chromosomes.

With no concrete end to the graft hybrid controversy in sight, Gates moved on. Just prior to taking up a series of lectureships in England, he made the acquaintance of William Bateson. Bateson struck up a correspondence with Gates in the spring of 1911, with the two biologists discussing de Vries and the tangled genetics of *Oenothera*. At one point, an unimpressed Bateson told Gates, "I expect de Vries had confused his homo- and hetero-zygotes. There are other such tangles in various parts of the work."[25] Bateson's inflexible stance on Mendelian genetics did not sit well with Gates. At a meeting of the Linnean Society of London in March 1914, the two clashed. "I was glad to get your letter," wrote Bateson in the aftermath, "for I could not understand why you 'challenged' me at the Linnean meeting." Bateson argued that the new concept of genotypic variation could be best explained as the consequence of changes in the germ cells, with Gates taking a broader view of variation and heredity. The two were quick to patch up their relationship, Bateson writing a few days later that he kept an open mind on variation and that "we had too many pretty speeches the other night."[26]

Back in the United States, Winkler faced opposition from a prominent and familiar face: William E. Castle, the same Harvard geneticist who had clashed with Guthrie and his graft hybrid chickens. In 1911 Castle was sent some apples from W. W. Clarke, from Nova Scotia. The fruit came from a shoot of Golden Russet that had been grafted onto the stock of the Boston Stripe variety. At the stem the apples resembled Golden Russet, before taking on the appearance of Boston Stripe halfway down. Castle, an avowed Mendelian and graft hybrid skeptic, could use chimerism to explain the strange hybridity of these apples. "Since the time when these 'freak' apples first came to my attention," wrote Castle in the *Journal of Heredity*, "the nature and origin of such plant creations has been investigated with brilliant success by Winkler and Baur in Germany." He did not, however, bother to mention Winkler's claim to have created true graft hybrids. Winkler, claimed Castle, had produced "plant-chimeras," the different cells of which "remain side by side but quite distinct in the same stem."[27] By adopting Winkler's findings on chimeras but ignoring his insistence on the existence of graft hybrids, Castle explained the strange appearance of fruit from Clarke's grafted trees without discussing any problematic challenges to genetics.

Initially, Winkler and his graft hybrid plants had excited much attention among the scientific community, particularly among foreign botanists. Campbell became an enthusiastic supporter, while Gates was sympathetic to mechanisms of heredity beyond Mendel but soon wavered in his support. Castle, as ever, was overtly opposed to graft hybrids. By effectively ignoring graft hybrids and focusing on chimeras, he could claim that there was no mixing of genetic material in grafted plants. Graft hybrids did not exist and could not challenge the validity of Mendelian genetics. As this consensus began to emerge, Bateson watched events unfold in Germany with acute interest. The rise of the chimera would offer him another opportunity to strike against the forces of neo-Lamarckism and the inheritance of acquired characters.

Bateson Encounters the Chimera

When William Bateson returned to Cambridge from the latest of his triumphant visits to the United States, his celebrity was not quite as pronounced in his homeland. Bateson had been isolated from the traditional biology of the university and labored to defend Mendelian genetics. As if to add to his woes, no sooner had he left the United States than Guthrie's graft hybrid chickens caused a lively controversy. Despite events across the Atlantic, however, back in England, things finally started to go Bateson's way. His fledgling research program in genetics had long received invaluable support from graduates of Newnham College, beginning with his sister Anna, who in 1890 used her botanical training to enter commercial horticulture. Edith Rebecca Saunders, another Newnham graduate and director of the Balfour Biological Laboratory in Cambridge, carried out vital hybridization experiments on seeds Bateson had collected from Italy in 1895. Bateson discussed Saunders's work at meetings of the Royal Horticultural Society, their collaboration allowing Bateson to recognize the implication of Mendel's paper for heredity upon its 1900 rediscovery.[28]

By the early years of the twentieth century, Bateson's Mendelian group had gathered more followers. Yet Bateson was in a personally precarious position. He had been appointed professor of biology at Cambridge, but the position had only enough funding for five years. In the meantime, there were no signs of any institutional recognition for genetics at Cambridge.[29] A new opportunity for Bateson arose in 1909, when he was offered a post as director at the newly founded John Innes Horticultural Institution. Although a move to the Norfolk-based institution would involve the loss of his professorship and its associated perks, Bateson jumped at the opportunity to hold a permanent post. The grounds of the John Innes and of Merton House, Bateson's new lodgings, were soon turned over to Mendelian hybridization experiments. He also used his new position to offer studentships to former members of his Cambridge research group, including Muriel Wheldale of Newnham College and William Backhouse of the School of Agriculture.[30] Around-the-clock access to plant breeding facilities also meant that Bateson had the ability to pursue some of the more complex facets of heredity, among them the graft hybrid phenomenon.

In January 1909, Bateson took a short vacation in Berlin. He visited museums, attended the theater, and repeatedly lost games of chess in the city's cafes. "The general gaiety of the place impresses me," he informed his wife, Beatrice, "quite as much as in Vienna, I should say." In what was probably less welcome news, Bateson confessed to her that "Money shews extreme tendency to dribble out of my pockets."[31] At some point during his stay, Bateson first met with Baur. Later that year, Bateson reached out to the German botanist in writing, recalling his "pleasant days in Berlin," where he had first had the "opportunity of making your acquaintance." Science, not pleasantries, was now on Bateson's mind. "It seems to me," he wrote, "that your experiments are some of the most interesting being made in the Mendelian field."[32] Bateson, a zoologist by training, was soon in regular contact with Baur, inquiring about the latest thinking on heredity among the botanical community. Bateson was frustrated by the backwardness of English botany when it came to heredity. "Can you put me onto any literature which gives the facts clearly and up to date?" he wrote to Baur in May 1910. "Our botanists here seem to know nothing particular about the question."[33]

Bateson was soon sufficiently invested in his relationship with Baur to update him about events at the John Innes Institution, modestly describing its five-acre grounds, greenhouses, small laboratory, and staff accommodation. Bateson also expressed the hope that the institution would one day reflect his interests in all aspects of genetics. "I consider that any Station of this kind," mused Bateson, "should aim at including work with animals as well as plants. You know well enough how closely the two subjects interweave, and each helps the other." He had also received some exciting news from Baur, who had managed to obtain some specimens of Winkler's chimeras and offered to dispatch them to England. On March 31, 1911, Bateson eagerly responded, "We shall be more than pleased to have the graft-hybrids," and suggested that he might exhibit them at a Royal Society soirée that May. Bateson was simultaneously conducting his own experiments on plant grafting at the John Innes Institution. "We are trying to make grafts too," he informed Baur, "but I don't know if we can do it right." He also urged his colleague to come to England, in what was only the first in a series of unsuccessful attempts to coax Baur into visiting him at Merton House.[34]

Baur might not have expressed much enthusiasm at the prospect of visiting the John Innes, but on April 13, 1911, Bateson received a visitor from Berlin of an entirely different nature. He informed Baur: "The plants have come safely to hand. The box was broken and I feared the contents might be injured, but they were in perfect order, I am thankful to say. We are delighted to have them." For the first time, Bateson had finally come face-to-face with a graft hybrid. Winkler's plants raised a whole series of theories and questions in his mind, which he quickly relayed to Baur:

> All this new development as to the extreme importance of the sub-epidermal layer seems to me to raise some very large questions. I do so wish you were here to talk them over! None of our botanists ever seem to have given the thing a thought.

> I am coming more and more to think that the sub-epidermal layer
> must really be the germ-tissue, already separated from the soma, like the
> germ-tissue of so many animals, according to the Weismannian view. I can't
> make out whether botanists are familiar with this possibility or not. No
> one seems to know anything about such things on the botanical side.[35]

Bateson also sought to confirm that the plants did not in any way suggest the
inheritance of acquired characters. Baur had argued that Winkler's plants
were not true hybrids but were composite organisms of two plants growing
side by side. Bateson used the remainder of his letter to continue with his
train of thought on the nature of chimeras, all the while bemoaning the
absence of Baur: "I [now] think of the sub-epidermal layer as consisting of
patches of the differentiated kinds, carried up the plant like the coloured
glass in a Venetian bottle. Would this not also agree with what you have discovered as to heredity of sectorial chimaeras? I do so wish you would give us
a week here. Now is the time, before the seeds are up. We should be delighted
to put you up at this house, and of course we are close to London. Why not
come? You are the only man who would understand the questions involved."[36]
Baur was not to be swayed, despite repeated pleas from Bateson to attend
the Royal Society soirée. Bateson kept abreast of the latest news from Germany. Although he did not read Winkler's original papers, he was still able
to follow the chimera controversy in some detail, likely thanks to Gates's
faithful reporting. "I have not seen the last Winkler paper," wrote Bateson
in late April, "but I understand that he has distinguished the cell-layers
in the hybrids by means of the chromosome numbers, as you suggested he
ought to do."[37] Despite the absence of Baur and the ongoing graft hybrid
controversy in Germany, the Royal Society soirée went well. "I showed the
Chimaeras at the Royal Society," Bateson reported on May 13, 1911, "where
several were much interested in them."[38] Following the success of the soirée,
Bateson traveled to Germany in 1912, stopping first in Berlin before meeting
with Baur in Potsdam. A brief walking vacation indicated that Bateson's
eagerness to spend quality time with Baur was probably misplaced. Bateson
was keen to discuss art and literature, subjects in which Baur had no interest.
In fact, Bateson discovered to some dismay, Baur's only real passion outside
of Mendelian genetics was his hatred of the French.[39]

Regardless of personal incompatibility, Bateson and Baur continued their
professional collaboration. Their next project was the organization of the
1915 Congress of Genetics in Copenhagen. Their Danish contact was the

geneticist Wilhelm Johannsen, who, unfortunately, was in poor health. "A letter at last from poor Johannsen," Bateson exclaimed in January 1913. "He has had a dreadful time and is evidently far from well. He is willing to try to organize a Congress for 1915 at Copenhagen, but I scarcely like to rely on him in the circumstances."[40] When this news reached Baur, however, the geneticist was less than sympathetic. Summarizing Baur's response for his own records, Bateson wrote that Baur "added [that] Johannsen was really better; that he, Johannsen, was always over anxious, and under careful of himself, and he returned Johannsen's letter."[41] Whatever the true nature of Johannsen's condition, an agreement was made to shift the Genetics Congress to Berlin. Plans for the next congress were even afoot, with Bateson touting Vienna as a possible venue. In a harbinger of controversies to come, Bateson remarked, "I wonder how the Kammerer discoveries will look at that remote date!"[42]

Global events, of course, derailed any thoughts of holding the Genetics Congress in Europe. Although the outbreak of the First World War in 1914 may have shelved scientific meetings, Bateson managed to maintain his copious flow of correspondence. In a letter to Signe Laura Amalia Nilsson-Ehle, wife of the Swedish botanist Herman Nilsson-Ehle, Bateson expressed his dismay at the thought that he might never be able to visit Berlin again. He even feared that international scientific relations would not recover in his lifetime. "I had hoped," Bateson confessed, "to see the world become more cosmopolitan, and nationality forgotten a little more as I grew older, but instead there has come this catastrophe." The nationalistic Baur had been gripped by war fever. "I had wanted to send Baur some friendly message last year," Bateson continued, "but I was told it was no use and that they are patriotism-mad. Do you hear from him? I should much like to know what he is doing."[43]

With his expert botanist cut off, Bateson pressed ahead with his research on plant chimeras. In December 1916 he published a short article in the *Journal of Genetics*, once again introducing the English-speaking world to "the work of Baur and Winkler," which he claimed had spurred universal interest in plant chimeras. "From his studies of variegated plants," Bateson explained, "Baur successfully interpreted these cases" of graft hybridization by demonstrating that graft hybrids consisted of different cell layers.[44] In Bateson's view, Baur had successfully uncovered the true nature of graft hybrids without any recourse to mechanisms of heredity beyond those exposed by Mendel. Moreover, chimeras could also be used to explain away some of those puzzling heredity phenomena that had so vexed him in the

United States. He continued: "The object of this note is to point out the fact that collateral evidence shows some unsuspected plants to be in reality of this nature, namely periclinal chimaeras, having an outer layer or cortex distinct in genetic composition from the inner core. My attention was called to the subject by reading a report of an address of Mr C. E. Pearson to the Horticultural Club (June 30, 1914), in which he stated that some Bouvardias, and the class of Pelargoniums known as 'Regals' did not come true from root-cuttings."[45] The bewilderment of Pearson when faced with flowers that grew differently depending on whether he planted a piece of root or a piece of stem could now be resolved. Bateson explained that Pearson had encountered periclinal chimeras, which could produce different types of flowers depending on the identity of the tissue they were grown from. He also reiterated the genic separation between tissues in a chimera. "Presumably," mused Bateson, chimeras "will include the various kinds of distinctions for which genetic factors are responsible." Usually, he concluded, the distinction between the different tissues of a chimera was restricted to the chloroplasts, with the exception of "the graft-hybrids."[46] Although Bateson continued to refer to *graft hybrids*, his publication made it clear that these plants were in fact chimeras, as described by Baur. Bateson did not stop to entertain the idea that genuine graft hybrids, with mixed genetics, might exist.

Baur and Bateson would reconnect after the war, the chimera controversy having essentially—at least in their eyes—been settled. Both biologists had new projects to pursue and scores to settle, particularly in the form of the troublesome neo-Lamarckian Paul Kammerer. Baur harnessed his revulsion toward the mentally ill into a drive for eugenic reform.[47] Bateson also came to believe that biology could be applied to human society, including the elimination of hereditary criminality.[48] His political and social priorities may have been somewhat altered by the events of the First World War. Bateson observed that although felonies may be exciting local events, they are largely insignificant compared to the damage done by warmongering politicians.[49] The end of hostilities also proved beneficial to his relationship with Baur. Bateson noted that Baur's correspondence in the 1920s struck a friendlier tone, suggesting that a "black dog had come off [Baur's] shoulders."[50]

Bateson's move to the John Innes Institution and his fascination with seemingly obscure phenomena such as plant chimeras has been portrayed as something of a misstep in his scientific career. During the later years of his career, he certainly struggled to keep abreast of the latest thinking in genetics, most notably taking an ill-fated stance against Thomas Hunt Morgan and chromosome theory.[51] From Bateson's perspective, however, a misstep was by no means immediately evident, even if he was ultimately lost among the details of plant genetics, development, and embryology. His years spent studying chimeras, including his 1911 exhibit at the Royal Society, addressed what, at the time, was a pressing issue. Winkler had claimed to have created true graft hybrids, causing ripples across the scientific world. A handful of Mendelians, most notably Baur, were engaged in an active intellectual battle to reject their existence. By labeling graft hybrids as chimeras and explaining the true nature of the latter to his scientific contemporaries, Bateson was simply seeing off yet another threat to Weismannism and Mendelism.

Burying a Bone of Contention

It took Bateson some five years after his chimera exhibit at the Royal Society to publish his findings on graft hybrids. When he finally did so, his paper on the subject appeared at the height of the First World War. Many of his fellow biologists had more pressing issues on their minds, especially those in active service such as Reginald Ruggles Gates. However, fresh work on the graft hybrid controversy did appear, including a study in the summer of 1913 by the botanist Margaret Hume. Like many of Bateson's female colleagues, Hume had also passed through the gates of Newnham College in Cambridge, graduating in 1910 and receiving the Bathurst Studentship to conduct advanced research. In 1911 she began to attend classes on plant ecology, which were then run by the ecologist Arthur Tansley.[52] A product of the Working Men's College in London, Tansley was an amalgam of ideas and contradictions. He would describe different ecological habitats in great detail during his Cambridge lectures before forgetting the names of common plants.[53]

In 1901 Tansley had launched his own botanical journal, the *New Phytologist*, to pursue his particular vision of biology. Vernon Blackman, a close friend of Tansley, discussed the rediscovery of Mendel and its implications for plant hybridization in one of the earliest issues of the journal. Blackman worked on fungi at the Natural History Museum in London but had also attended the Botany School at Cambridge and even worked in Strasburger's laboratory in Bonn.[54] "That Mendel's law is of the greatest importance in relation to the general laws of heredity," wrote Blackman in 1902, "and to our conception of the relation of characters in the organism there can be no doubt." Blackman proclaimed that Mendel's laws had created a new perspective on hybridization, quoting Bateson's vision of the organism as "a complex of characters of which some at least are dissociable and capable of being replaced by others."[55] Tansley also possessed a firm grasp of Mendelian principles. An anonymous correspondent wrote to him in 1912 to congratulate him on his mastery of genetics: "I have read your paper through carefully, and I think it is most excellent. It explains more clearly than anything I have previously read the lines along which Mendelian research is moving. This research is getting results which seem to me to appeal to commonsense far more than any old hypothesis."[56]

Although Tansley and Blackman embraced many aspects of Mendelian genetics, graft hybrids also appeared in the *New Phytologist*. In 1911 the journal carried an article by a regular reviewer who recapped events in Germany. Winkler's plants, the reviewer explained, had eventually been outed as periclinal chimeras by Baur. The one exception to this rule was the *Solanum darwinianum*. The plant possessed twenty-four chromosomes, which, the reviewer explained, was indicative of hybridization. "Further investigations into the origin of *S. Darwinianum*," they explained, "will be awaited with great interest, for at the present time it alone affords any positive indication that such cell-fusions may take place, giving rise to cells, and ultimately to tissues, which are essentially of a hybrid nature."[57] The *Solanum darwinianum*, with its unusual chromosome count, was fast becoming the go-to example for defenders of the graft hybrid. This trend would further accelerate when Hume eliminated another line of argument for the existence of graft hybrids.

If Hume was not already a Mendelian by the time she encountered Tansley in 1911, she was certainly skeptical when it came to the existence of graft hybrids. Her entrance to the graft hybrid debate was published in the *New Phytologist* in 1913. While stationed at the Botany School in Cambridge, Hume decided to investigate a claim that minuscule threads could connect different cells. These threads, ran the argument, could act as a genetic connection between two grafted plants. Strasburger, now well known as a graft hybrid debunker, had already conducted his own investigation into the existence of these threads. His findings were inconclusive. Although the mysterious threads did exist and did connect genetically distinct cells, Strasburger had found no evidence that they carried genes between cells. "The point where a graft is united with the stock," noted Hume, "is a region where tissues of different genetic origin are in living continuity, but the differentiation between the tissues of stock and scion cannot be easy." Knowing whether these different tissues could exchange genes in the absence of cell fusion was of huge importance in the wake of Winkler's "artificial production of graft hybrids." Hume was openly unconvinced by Winkler's findings, seeing "no reason to doubt that Baur is right" in labeling Winkler's plants as periclinal chimeras.[58]

Hume began her investigation with a specimen of *Cytisus adami*, obtained from the Royal Botanic Gardens, Kew. This plant had pedigree, having been labeled a graft hybrid by Charles Darwin. It had first been examined under the microscope in the late nineteenth century by John Muirhead Macfarlane, a Scottish botanist at the University of Edinburgh. In the summer of 1891, Macfarlane had discussed his findings at a meeting of the Royal Society of Edinburgh. Under the microscope, he claimed, the plant's fibrous tissue appeared to be of an intermediate type, resembling a hybrid of its grafted "parents." Macfarlane told his Edinburgh audience that the plant left him with no choice but to admit that "cell unions may be effected without intervention of sexual elements."[59] Thanks to his skill with a microscope, Macfarlane was soon offered a comfortable professorship at the University of Pennsylvania. He did not discuss graft hybrids at the 1899 meeting of the Royal Horticultural Society and avoided all future controversy around them.[60] Subsequent examinations of the *Cytisus adami* by Strasburger would cast doubt on its hybrid nature. Even Winkler acknowledged that the plant was probably a chimera.

Given the controversy surrounding every technical detail in the graft hybrid debate, Hume meticulously described her methodology. She used an established method for preparing her microscope slides, which included fixing her specimens in an iodine solution and staining them with safranin. Hume chose to work on the flower of *Cytisus adami*, where its intermediate characteristics were most prominent. Under the microscope she saw "deep pits [that] connect the cells of the epidermis with those of the underlying layer," confirming earlier observations made by German botanist Johannes Buder. Hume went one step further and examined other sections of her plant. Her investigations revealed that the "pits and their fillings do not differ in appearance from those which connect the cortical cells of *Cytisus Adami* with one another."[61] The fact that these connections existed throughout the *Cytisus adami* indicated that they were probably a physiological structure, rather than a means of genetic exchange.

With this nineteenth-century graft hybrid out of the way, Hume moved on to more contemporary matters. By approaching the John Innes Horticultural Institution, Hume was able to obtain material from some of Winkler's graft hybrids, including the *Solanum tubingense*. This material may well have come from the plants originally supplied by Baur for Bateson to exhibit at the Royal Society, or from cuttings of those specimens. Hume's description of the plant, as having the skin of a tomato plant overlaying black nightshade, matched that given by Campbell when he had encountered it in the Munich botanic gardens. Hume began by examining the stem of the plant, which resembled tomato but grew leaves of nightshade. She found that "the connecting threads between the epidermis and the underlying layer of cells, are demonstrated very easily." Once again, Hume found nothing unique about these connections. "The threads connecting the epidermal cells with those of the layer below," she recorded, "correspond exactly with those connecting together neighbouring epidermal cells or neighbouring cortical cells."[62] Another of Winkler's graft hybrids, the *Solanum koelreuterianum*, did not appear to possess any connecting threads at all.

In Hume's view, the lessons to be drawn from her physiological examination of different graft hybrids were clear. The presence of connecting threads did not imply that genetic exchange between different tissues had taken place. These threads were structural in their function, holding no relationship to the similar-looking spindle-fibers seen in cells during division. Hume noted that connecting threads appeared in every part of a plant, arguing that those "joining cells of genetically related and cells of genetically unrelated tissues are to all appearances entirely the same." As these threads were not unique to the graft junction, it was highly unlikely that they played any role in the transmission of genetic information. She concluded: "The principal conclusion to be drawn is, that if Baur's hypothesis is true, and there is every reason to believe that it is, and that graft hybrids really are periclinal chimaeras, then there is no doubt that genetically unrelated tissues can be joined by connecting threads. The threads therefore arise secondarily, since it is to be supposed that the naked cytoplasm of the two components does not come into contact. At any rate it is clear that the threads cannot have arisen from spindle-fibres, since no nuclei of the two components have ever been sisters."[63]

Following her intervention into the graft hybrid debate, Hume departed England in 1913 for a lectureship in botany at the South African College in Cape Town. In one sense, her paper was just another in what was becoming a very long line of published attacks on Winkler. In another sense, it was remarkable that a rather technical controversy over graft hybrids and chimeras had engaged an international community of the biologically inclined. This fact, however, would have been of small comfort to Winkler. With most of his precious plants outed as chimeras, all he had to fall back on was his *Solanum darwinianum*. The plant had not yet been examined by Hume, or any other botanist. It also had a promising chromosome count in its favor.

Still, across the Atlantic, the American Paul Bowman Popenoe, editor of the *Journal of Heredity*, suggested that the sun was setting on the graft hybrid. Chimeras, Popenoe remarked, "were first reported nearly three centuries ago, and doubtless existed centuries earlier, but unnoticed. It was not until lately, however, that their production under experimental conditions, and the study of their cells under the microscope, made it possible for botanists to understand exactly what they were. Now that we know the trick, they can be produced by anyone with patience, and the mystery surrounding the so-called 'graft hybrid,' a bone of contention among horticulturists for several centuries, has vanished."[64] The only exception to this rule, noted Popenoe, was the *Solanum darwinianum*. Its distinctive number of chromosomes, if confirmed, would grant it "a strong claim to be considered a real graft-hybrid, the first one ever known" and "the only genuine, out-and-out graft-hybrid in the world."[65]

The graft hybrid controversy in Germany, now closely monitored by biologists in Europe and North America, looked set to enter its final phase. Would Winkler's chromosome count of the *Solanum darwinianum* be confirmed by others? Would further investigation into the nature of the plant reveal it to be a true graft hybrid, or yet another chimera? Unfortunately, these questions would never be answered. During the First World War, the *Solanum darwinianum* was somehow lost. Whether it was misplaced or simply died is unclear. Regardless, its loss was disastrous for Winkler. As Charles F. Swingle of the US Bureau of Plant Industry would later conclude, the plant "would always remain a puzzle," in part because of its wartime disappearance, in part because of "the impossibility of ever obtaining the identical graft hybrid again and unmistakingly recognizing it as such."[66]

The demise of *Solanum darwinianum* was a serious blow to Winkler's ambition to create a graft hybrid under experimental conditions. Despite the loss, he persevered with his research. Working with dehydrated tissue of *Solanum* plants, Winkler was able to artificially combine them to produce buds with four sets of chromosomes. He also attempted to recreate some of his grafted plants, attaching pieces of nightshade to tomato stocks. Once again, chimeras emerged from the intersection of the two plants. In one case, Winkler was able to obtain a branch of pure tomato with four sets of chromosomes. In another, he acquired a nightshade plant with over a hundred chromosomes in its cells. Though interesting, boosting the number of chromosomes in a single species of plant did not lend support to Winkler's claim that grafting could allow plants to hybridize. His wartime experiments, published in 1916, were

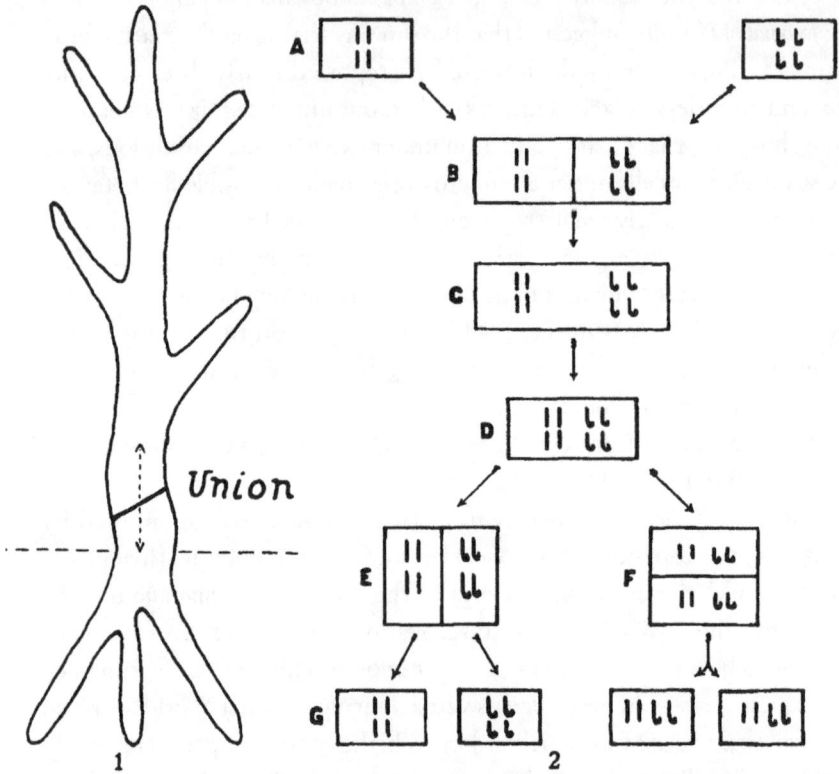

FIGURE 5. Cell fusion graft hybrid. This image of how a graft hybrid could hypothetically be created by cell fusion was produced by Charles F. Swingle of the United States Bureau of Plant Industry. It depicts chromosomes from the two joined plants (A) merging through the graft junction (B to D). Theoretically, these chromosomes could combine and be inherited by the offspring of the grafted plants (F). Charles F. Swingle, "Graft Hybrids in Plants," *Journal of Heredity* 18, no. 2 (1927): 77. Reproduced by permission from Oxford University Press.

of more interest to traditional plant breeders with their varietal crosses and backcrosses.[67] With the end of the war, Winkler rode a wave of renewed interest in cytoplasmic inheritance in Germany to deny that hereditary material was confined to chromosomes. At the 1923 meeting of the German Genetics Society, Winkler argued that some traits existed in the cytoplasm and were carried to the cells by "plasmagenes." Though widely cited, his arguments were not widely accepted or backed by hard evidence.[68]

By the end of the First World War, botanists and plant physiologists increasingly identified graft hybrids as chimeras. Hume's investigations played a role in this developing consensus. Her examination of spindle-fibers undermined the argument that grafted plants could subtly exchange genetic material. Cell fusion, as confirmed by chromosome counts, was now the only accepted means of defining whether an organism was a graft hybrid or a chimera (for a hypothetical model of such a graft hybrid, see fig. 5). The battle between graft hybrids and chimeras had been fought on an international stage on many grounds: natural-history-style travels and observation, physiological studies, and chromosome counts. Yet even advocates of graft hybridization, like the American botanist Campbell, conceded that observation alone had little to offer. When dissected and placed under Hume's microscope, grafted plants invariably revealed themselves to be chimeras. This left graft hybridizers searching for evidence of cell fusion using chromosome counts. The only plant created in Germany whose chromosome numbers indicated that cell fusion had occurred was Winkler's *Solanum darwinianum*. Its mysterious disappearance during the war was therefore even more devastating to the graft hybrid cause, as all other lines of evidence had simply fallen away.

An Ancient Myth

The graft hybrid may have had its supporters, but by the late 1900s a new hypothesis had arisen to challenge it. Winkler had originally conceived of the chimera as a means of explaining the appearance of some of his less convincing graft hybrids. With the intervention of Baur, Hume, and other botanists, it became a new means of refuting the countless botanical curiosities of previous centuries. Though fascinating in their own right, chimeras had no bearing on questions of heredity or genetics. When German geneticists returned to international science at the end of the First World War, there would be no reconciliation between opposing camps on the graft hybrid question. Although some members of the biological sciences in Germany remained sympathetic to the idea of cytoplasmic inheritance, Baur and his chimeras had largely won out. Popenoe seemed somewhat relieved that chimeras had superseded the graft hybrid. "To the old botanists," he wrote in 1914, "the existence of supposed graft-hybrids seems at times to have been rather resented—they interfered so much with theories about heredity! Fortunately we have passed that stage."[69]

The rise of the chimera did not only serve to reinforce Mendelians countering what they saw as neo-Lamarckian beliefs. Chimeras also changed minds about the existence of graft hybrids. Among those whose minds were changed was the American botanist and geneticist George Harrison Shull, whom we encountered corresponding with Davenport over photographs of a tomato graft hybrid in 1910. In 1906 Shull had argued that graft hybrids could be created through cell fusion. Responding to the investigations of a German botanist into some unusual grafted plants at Bronvaux, France, he declared that as "the investigation shows that neither stock nor scion is itself of hybrid origin . . . there can be no reasonable doubt that these are true graft-hybrids."[70] This, however, was before the clash of Winkler and Baur. A little under a decade later, in 1915, Shull addressed the nature of the Bronvaux plants once more. His perspective had changed radically. "The view of Baur is sustained throughout," wrote Shull, demonstrating "that these two 'graft hybrids' consist of a core of *Crataegus* tissue overlaid by a mantle of *Mespilus*."[71] Confronted with Baur's reinterpretation of graft hybrids, Shull reversed his position and declared the Bronvaux plants to be chimeras. Only a few years earlier, the United States had been an arena where different theories of heredity and methods of plant breeding had existed side by side. Times were now changing—within American universities, at least. When Richard Goldschmidt, a German geneticist, visited the United States in the 1920s, he was upset to find that the Morgan school of genetics had become focused on "pure Mendelism," to the exclusion of physiology and embryology.[72] Graft hybrids had never been embraced by geneticists. Now they had lost ground among breeders and universities as well.

Only a few months before the First World War descended on Europe, Scottish plant physiologist Macgregor Skene stood surveying the carnage of this intellectual battle over the nature of heredity. From the legions of intermediate forms produced by hybridization over the centuries, wrote Skene, the occasional appearance of a graft hybrid had drawn "a considerable amount of attention" from botanists. It was only in recent years that an "experimental solution" to the graft hybrid question had emerged. "When Prof. Winkler commenced his investigations some seven years ago our knowledge of the origin of these curious plants was sadly indefinite," recorded Skene. "That they had arisen by grafting seemed improbable; that they possessed properties seen in no other hybrids made a sexual origin equally doubtful."[73] He concluded that, with the benefit of hindsight, "Prof. Baur's theory proved to be correct. The hybrids are indeed plants in which a core of one parent is enclosed in a skin of the other." Graft hybrids were not true hybrids, but rather "the materialisation of a very ancient myth."[74] As Europe emerged from the First World War, most biologists would have agreed with him. Graft hybridizers, however, ranging from respectable botanists to notorious neo-Lamarckians, would not be so easily silenced.

3

Creed in the Place of Science

IN 1924 AN ANONYMOUS CORRESPONDENT WROTE TO THE PRESTIGIOUS scientific journal *Nature* to resurrect what many believed to be a discredited hypothesis. Their short note described the experiments of Lucien Louis Daniel, professor of agricultural botany at the Lycée de Rennes in France, who had grafted artichokes onto sunflowers. Unlike Charles Darwin and Hans Winkler, however, Daniel had managed to get seeds from his grafted plants. These seeds in turn grew into plants with different colors, weights, and tubers, which suggested the creation of new varieties through grafting. "This is claimed," explained the correspondent, "as a case of the inheritance of a character acquired by grafting, and it is difficult to see how this interpretation can be denied."[1]

This dramatic assertion attracted the attention of Wilhelm Johannsen, the Danish geneticist whom we last encountered anxious over the organization of the planned 1915 Genetics Congress. Johannsen responded in *Nature*, noting that artichoke seed was naturally highly heterozygous, meaning that it inherited different forms of genes from its parents. The varieties created by Daniel, argued Johannsen, were evidence only of the natural variability of artichoke, not the inheritance of acquired characters. He went on: "Grafting very often causes the scion to produce flowers and seed; it seems very natural that the transplanted artichoke was influenced in that way—a physiological action of purely phenotypical nature without the slightest influence upon the genotypical constitution of the scion; the differences between the individuals of the progeny are here without doubt a consequence of the heterozygous nature of the scion. I must confess that the great bulk of indications claimed as cases of the alleged inheritance of acquired characters show an astonishing lack as to critical judgment—creed in the place of science."[2]

Despite Johannsen's claim that belief in graft hybridization showed a lack of "critical judgement," the chimera controversy in prewar Germany had not entirely removed the graft hybrid from the realm of respectable science. For one, the mystery and controversy which had swelled around the experiments of Charles Claude Guthrie and Hans Winkler had never been satisfactorily resolved. Their graft hybrids had been lost and few efforts were made to replicate them. Daniel, seemingly unperturbed by any theoretical arguments, had spent his long and illustrious career producing new varieties of fruits and vegetables for French horticulturalists through grafting. Developments in genetics also continued apace, with Thomas Hunt Morgan's chromosome theory giving a physical location for the hereditary unit.[3] Graft hybridizers now had to incorporate these findings into their practice and interpretations of past experiments.

The chimera controversy in prewar Germany had not driven the graft hybrid out of science. It had, however, isolated graft hybridizers and would soon drive a wedge between them and the ordinary horticulturalists and gardeners who would have been among their strongest supporters. On the plus side, the growing complexity of genetic science introduced fresh ambiguity and nuance into the study of heredity: an environment in which a diplomatic graft hybridizer could still operate. Against this backdrop of unresolved arguments and contemporary discoveries, a new generation of graft hybridizers flourished in the ambiguities of the interwar period. Three of these graft hybridizers, each in a different metropolitan setting, deployed separate strategies to support their increasingly problematic beliefs. At the Viennese Vivarium, or Institute for Experimental Biology, the flamboyant zoologist Paul Kammerer produced salamander graft hybrids to support his wider Lamarckian vision. Frederick Ernest Weiss, the botanist son of German émigrés, gave his support to graft hybridization within the scientific and civic societies of Manchester, England. Meanwhile, William Neilson Jones, head of the Botany Department at Bedford College, London, lent his support to graft hybrids from inside the higher education system.

The trio of Kammerer, Weiss, and Neilson Jones also embody three different styles of communication pursued by graft hybridizers in the interwar period. Kammerer favored confrontation, conducting his own hybridization experiments, and taking his results public. Weiss preferred reconciliation, embracing Mendelian genetics and graft hybridization simultaneously. Neilson Jones chose appeasement, supporting graft hybridization as a purely theoretical concept and distancing himself from its supporters among farmers and breeders. Unfortunately for the trio, none of these strategies allowed the graft hybrid to regain traction in Western biology or agriculture. Kammerer was denounced as a scientific fraud. Weiss largely confined his musings on graft hybridization to the civic societies of Manchester before revising his beliefs in the 1940s. Neilson Jones published his comprehensive *Plant Chimaeras and Graft Hybrids*, based on a lecture series at the University of London, in 1934. When it was reissued in 1969, however, it was described by one reviewer as haunted by the ghost of the graft hybrid hypothesis.[4] As the twentieth century progressed, Western biologists increasingly came to associate graft hybridization with the revolutionary politics and controversial science of the Soviet Union.

Mental Independence from Mendel

In 1923 Paul Kammerer arrived in Cambridge, England, to public and scientific acclaim. Following a lecture at the Natural History Society of Cambridge, the university's professor of zoology reported that "Kammerer begins where Darwin left off." His colleague the professor of biology G. H. F. Nuttall went further and declared that Kammerer "has made perhaps the greatest biological discovery of the century."[5] These remarks, particularly the comparison to Darwin, were eagerly seized upon by newspapers. According to his admirers, Kammerer had succeeded in demonstrating that characters acquired during the life of an organism could be inherited by future generations. By altering the environment in which he raised animals, particularly amphibians, he encouraged salamanders to change their color and midwife toads to develop special pads on their forelimbs for mating on land. William Bateson, of course, was not pleased. He had first encountered Kammerer over a decade earlier and had already formed an unfavorable impression of him. The Viennese zoologist had come "uncommonly near showing that an acquired adaption is transmitted." Bateson did not shirk from his feelings. "I don't like it," he wrote, "and shall not give in till no doubt remains."[6]

Some of Bateson's misgivings came from the fact that Kammerer's path into science from the world of music had been somewhat unusual. Born into a middle-class Viennese family with Jewish ancestry on his mother's side, Kammerer quickly developed a love of music and of caring for plants and animals. He would graduate from the University of Vienna and the Vienna Music Conservatory after studying biology, music, and philosophy.[7] Kammerer started his scientific career in earnest at the Vivarium, a former zoo and aquarium. Bought and refurbished by a group of wealthy scientists, including the brothers Hans and Karl Przibram, the Vivarium was revived as the Institute for Experimental Biology in 1902. Kammerer arrived shortly thereafter. The Vivarium stood out from other biological institutes at the time by virtue of its close-knit community and advanced environmental-control technology. Outside of the traditional and catholicized confines of Austrian academia, the Vivarium attracted members of affluent families with interests in science and the arts. It also possessed a strong connection with Vienna's Jewish community. Inside the walls of the Vivarium was an array of artificial habitats, with precision control over such factors as temperature and lighting.[8] Members of the Vivarium could use these tools to observe how changes to the environment affected the organisms within it.

Hired by Hans Przibram as a laboratory assistant, Kammerer conducted his doctoral research on two species of salamander. The first, a lowland variety, carried large numbers of eggs, which it deposited in water to hatch as tadpoles. The second, an alpine variety, carried fewer eggs, which hatched into larger, land-based forms. By harnessing the technological wizardry of the Vivarium, Kammerer was able to create dry or wet conditions for each salamander. By the end of his experiment, they had effectively switched roles, with the lowland species carrying fewer eggs, which delivered tadpoles adapted for life on land. The reverse was true for its alpine counterpart.[9] With his dissertation in hand in 1904, Kammerer moved on to more diverse and lengthy experiments. Over the coming years, he declared having produced abundant evidence of the inheritance of acquired characters. One of the most famous examples displayed by Kammerer was the midwife toad, which acquired "nuptial pads" on its forelegs when encouraged to mate on dry land. Kammerer also amputated the siphons of sea squirts, which grew back longer with every succeeding generation. He even claimed that his modified sea squirts could regenerate their sex cells from bodily tissue. "Through this restitution alone," he wrote, "a whole theory is overthrown—Weismann's theory of the continuity of germ plasm."[10]

After several years of such experiments, Kammerer, like biologists across the world, was hit by the impact of the First World War. His wartime service consisted of acting as a military censor for the Austro-Hungarian Empire. This work kept him away from the Vivarium, where many of his experimental organisms subsequently perished from neglect. Kammerer had never taken the time to systematically collect and preserve specimens, preferring instead to show off his living animals at the Vivarium to visiting scientists. The institute itself was not equipped for photography, with Przibram relying on his own artistic talent in order to draw specimens. Kammerer, who was a musician, not an illustrator, was forced to turn to a local studio to photograph his animals. With the Viennese studio "unaccustomed to having glistening, wet amphibians sit for portraits," the resulting images were almost always of poor quality or had to be later retouched.[11] Kammerer lashed out when technical difficulties marred the publication of his work. In 1913 he clashed with Erwin Baur, not over Mendelian principles or acquired characters but over what Kammerer considered a poor reproduction of his photographs in one of Baur's books. Baur retaliated by noting that Kammerer's photographs had preexisting shortcomings and had been retouched. Baur did not consider modifications to photographs problematic but did argue that they could not serve as "documentary evidence" in the same way original images could.[12]

Despite media appearances, by the time Kammerer visited Cambridge in 1923 his scientific career was in jeopardy. He had been turned down for a professorship in 1919 and left the Vivarium in 1921 to make a living from his popular writing and lectures. A British colleague and sympathizer, Ernest MacBride, professor of zoology at Imperial College London, would later depict interwar Vienna as ravaged by socialism. Under these circumstances, the University of Vienna could pay only "starvation salaries," hence Kammerer's turn to "journalism and popular lecturing."[13] In reality, Kammerer's application for professorship was undermined when he published a book on coincidences, which the committee considered pseudoscience.[14] In the wake of this affair, Kammerer did take the time to write up his previous findings in English. One of the experiments he had conducted before the war involved salamanders, the subject of his doctoral dissertation. Describing his experiments in *Nature* in 1923, Kammerer explained how black-and-yellow salamanders changed their markings and color in direct response to their environment. "If the young animals are kept on a black background," he wrote, "they lose much of their yellow marking and, after some years, appear mainly black." Subsequent generations of salamanders exhibited still greater change. If kept on a black background, they kept their natural spotted pattern. If, however, the fresh generation of salamanders grew up on a yellow background, these spots fused together to form a band. Kammerer argued that, through modifications to the environment, he had created an artificial form of striped salamander.[15]

With salamanders in hand, Kammerer turned to a new means of demonstrating the inheritance of acquired characters: graft hybridization. Three varieties of salamander were involved in Kammerer's efforts at ovarian transplantation. The first was the naturally occurring spotted type. The second was also a naturally occurring type but with stripes, imported from the Harz Mountains in northern Germany. Third, and finally, was the artificially striped salamander created by Kammerer at the Vivarium. As a precursor to the transplantation experiments, Kammerer had already crossed these varieties and found that Mendelian ratios appeared in crosses of the first two naturally occurring salamanders. Crosses with the artificially striped salamanders, however, did not result in offspring that conformed to Mendelian ratios. Kammerer found a similar outcome when he conducted ovarian transplantations. When a spotted male was bred with a naturally striped female, the latter with ovaries grafted from a spotted female, the offspring were all spotted. So far, so Mendelian. When Kammerer turned to his artificially striped salamanders, however, he obtained very different results. When a spotted male was bred with an artificially striped female, with ovaries from a spotted female, the offspring had an intermediate appearance, with spots arranged in lines (fig. 6).[16] This suggested that the body of the artificially striped salamander mother was exerting an influence on the appearance of her offspring.

To explain these results, Kammerer theorized that artificial stripes on salamanders exerted a greater influence than natural stripes or spots on heredity. "If one succeeds in preserving one's mental independence from Mendel's scheme," he wrote, "one will notice that, as far as the new striped race is concerned, the whole progeny—the children and also all the grandchildren—inherit the striped design." Recent changes to an organism driven by alterations to its environment, Kammerer thought, could overwhelm established characteristics. How did such a mechanism work? Kammerer was hazy on the details. He claimed that a character newly acquired by environmental changes provoked a kind of "morphogenetic irritation" in the organism. This irritation "possesses a great radiating power," occasionally powerful enough to reach the germplasm and alter inheritance. Kammerer further suggested that the longer a new characteristic remained in the body, the less potent its influence on heredity.[17]

FIGURE 6. Transplantation in salamanders. The Viennese zoologist Paul Kammerer attempted to demonstrate how changes to the environment could induce heritable changes by using grafting. Just as Charles Claude Guthrie had done with chickens, Kammerer transplanted ovaries between different varieties of salamander. He claimed that some of the salamanders' offspring displayed characteristics from both the birth mother and the donor of the implanted ovaries. Kammerer, *The Inheritance of Acquired Characteristics* (New York: Boni and Liveright, 1924), plate facing p. 100.

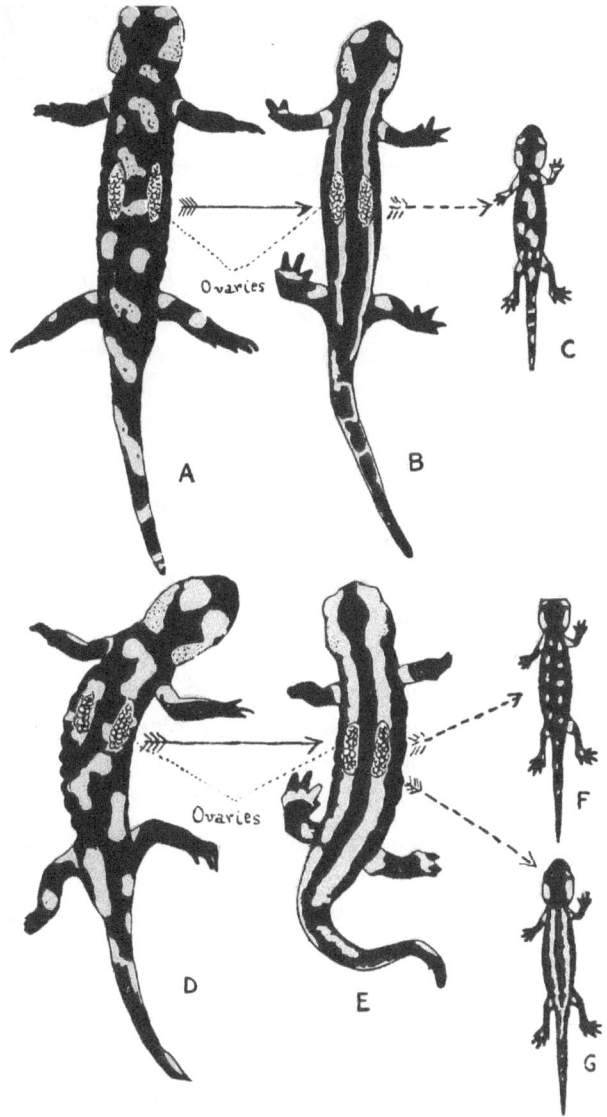

Anticipating a possible objection from August Weismann's acolytes, Kammerer argued that his earlier breeding experiments with salamanders had demonstrated the independence and function of the spotted salamander ovaries he later transplanted.[18]

What importance did the salamander transplantations hold in Kammerer's broader experimental program? In his 1923 article Kammerer held that they had played a fundamental role in convincing him of the reality of the inheritance of acquired characters. "These experiments on ovarian transplantation," he proclaimed, "first led me to consider the possibility of the true inheritance of somatic characters." Of course, Kammerer was not the first to perform such experiments, a fact he was aware of. Initially, he was dismissive of the transplantation experiments carried out on poultry by the American physiologist Charles Claude Guthrie. In Kammerer's retelling of the controversy, "portions of the original ovaries may have been left behind in the foster-mother" by Guthrie. When these chickens "were afterwards tested by [Charles] Davenport," their original ovaries were found to have regenerated and taken over from the transplanted ones. There were three reasons Kammerer was skeptical of his American contemporary. Firstly, he had encountered only the public rebuttal of Guthrie by Davenport, published in the *Journal of Morphology* in 1911, missing other articles with different perspectives on the controversy. Secondly, the private correspondence between the two Americans, in which Guthrie defended his experiments and left Davenport somewhat more receptive to his ideas, was of course unavailable to Kammerer. Thirdly, the Viennese zoologist thought that the entangled anatomy of the reproductive system in chickens was problematic, making salamanders a more suitable experimental animal. "Thanks to its enclosing membrane," he explained, "the ovary of the Salamandra can be removed from the surrounding tissue as a whole." It was therefore "impossible that any remnants could have been left behind and that the descendants were derived from these remnants regenerated."[19]

The other great graft hybrid controversy of the first half of the twentieth century also attracted Kammerer's attention. Winkler, a close geographical contemporary, had claimed to have created plant graft hybrids in the years leading up to the First World War. Kammerer noted that although grafted plants of "intermediate forms" were well known to "botanists and practical gardeners," most of these were not graft hybrids but chimeras.[20] Winkler's famous *Solanum darwinianum*, however, was a possible exception. Kammerer believed that a true botanical graft hybrid could occur only if two cells from each plant fused together in the same way sex cells did. The chromosome count of the *Solanum darwinianum* seemed to indicate such an event had occurred. This count, claimed Kammerer, "makes the assumption of an amalgamation, a copulation of the tissues grafted onto each other, plausible." He saw this as another blow against Weismann, as graft hybridization encroached on territory that was once exclusive to the germplasm. Grafting allowed the cells of the body to "prove their ability to amalgamate for the purposes of fertilization and so give start to new developments, the inception of a new individual."[21]

Despite the importance Kammerer attached to graft hybridization, his opinions on the technique and transplantation experiments on salamanders made little impact on either the scientific or the public sphere. Bateson, who was one of Kammerer's most vocal and consistent critics, had been aware of the latter's salamander graft hybrids for some time. In his 1913 *Problems of Genetics*, Bateson addressed the experiments on salamanders but reserved his judgment on their validity: "In continuation of the experiments on the colour of *S. maculosa* Kammerer publishes an account of elaborate experiments in grafting ovaries of the various forms, modified and unmodified, into each other, and describes the offspring which followed. Before pursuing this part of the inquiry I am disposed to wait until the earlier steps have been made much more secure than they yet are."[22] Bateson expanded upon his position in *Nature* in 1919, observing that salamanders that had similar patterns to those obtained by Kammerer were available from animal dealers. The midwife toad, whose nuptial pads "did not exist in Nature," were therefore of greater interest.[23] One defender of Kammerer would later suggest that Bateson's remarks held the veiled implication of fraud, as the salamanders produced through breeding and grafting could simply have been purchased.[24] A more likely explanation, however, is that Bateson took a keen personal interest in the midwife toad. The sudden emergence of a new characteristic such as the nuptial pad would support Bateson's view of Darwinian evolution as discontinuous rather than gradual.[25] Whatever the case, Bateson's intervention on the midwife toad would be one of his last sallies into scientific controversy. He was in failing health and died in 1926.

A far more serious critique of Kammerer's salamander transplantation experiments came not from a Mendelian but from a fellow neo-Lamarckian. Joseph Thomas Cunningham was a marine biologist who had claimed that the production of pigment in flatfish in response to light was evidence of Lamarckian adaptation.[26] In May 1923 Cunningham outlined his misgivings about Kammerer's transplantation experiments. The claim that salamander ovaries, unlike those of poultry, were surrounded by a membrane and could therefore be easily removed from the body, was of particular concern. "I have never heard hitherto of the existence of an enclosed ovary in any amphibian," wrote Cunningham. In fact, he understood the ovaries in amphibians to be as exposed to the surrounding tissue as their counterparts in birds. In both groups of animals, ova were able to pass through this surface. If it were the case that amphibian ovaries were not enclosed, mused Cunningham, "it would be very difficult to understand how ovarian transplantation could be carried out as in Kammerer's experiments." He did not underestimate the importance of the experiments, which, if accurate, would lead to "the extraordinary conclusion" that "the artificially striped soma makes the ova derived from a naturally spotted female behave as though they came from a striped female." However, Cunningham concluded, corroboration was required before biologists could accept such a monumental finding.[27]

The wider impact of Kammerer's graft hybridization experiments was further reduced by their irrelevance to agriculture. In a popular 1924 book explaining his findings for English readers, Kammerer confessed that "I am neither a practical breeder nor an expert farmer; about the raising of domesticated animals I know just about as little as about the raising of crops." He noted that his work on heredity had been conducted on the "lower animals," but insisted that its lessons were still applicable to the breeding of both domesticated animals and humans.[28] By contrast, Guthrie had worked with poultry on his family farm and appealed directly to his agricultural peers at the American Breeders' Association. He had also taken pains to depict graft hybridization as a potential tool to add to the armory of breeders. Kammerer, on the other hand, railed against the contemporary paradigm of farmers selecting the best animals to breed from. He instead insisted that the "training" of animals to perform certain tasks, from milking to racing, could be passed on from generation to generation.[29] Kammerer's intervention in the world of agriculture is best described as short and vague. Unlike Guthrie, he did not appeal to a community that had traditionally been open to the possibility of graft hybridization.

In 1926 a midwife toad specimen belonging to Kammerer was revealed to have been injected with ink at the site of its supposed nuptial pads, raising accusations that he was engaged in scientific fraud. Six weeks after the scandal broke, Kammerer took his own life. His defenders denied that Kammerer had committed fraud, with some even raising the possibility of a conspiracy against him.[30] The midwife toad threw suspicion on all Kammerer's biological endeavors, with the supposed fraud raised as means of dismissing all of his experiments throughout the twentieth century.[31] Yet even if we imagine a world without the midwife toad controversy, it is hard to imagine Kammerer's attempts at graft hybridization making much impact. Even the neo-Lamarckian Cunningham criticized his findings, suggesting that Kammerer might have failed to completely remove the ovaries of his salamanders prior to transplantation. When confronted with the same critique, Guthrie was at least able to defend his work based on his surgical expertise. Moreover, Guthrie's experiments on chickens had also appealed to open-minded agriculturalists of the early twentieth-century United States. On both counts, the unorthodox and metropolitan Kammerer could not say the same.

Assimilation and Interwar Life

"What, then, is this life?" asked Frederick Ernest Weiss, newly appointed professor of botany at Owens College in Manchester, England, to a meeting of the Manchester Microscopical Society in 1898. Weiss was not in the grip of an existential crisis, but was instead concerned with the fundamental question of what, if anything, differentiated organic life from inorganic matter. Growth was one common characteristic assigned to living things. Weiss noted, however, that "some inorganic bodies are capable of growth. Crystals, for instance, are said to grow, for all large crystals are formed from pre-existing smaller ones."[32] What made life different from crystals was that crystals required more of the same particles of which they were made to grow larger. To grow a crystal of rock salt, for example, it would need to be suspended in a salt solution. Life was different. Living things, Weiss argued, could consume almost anything in their environment. Plants could extract carbon dioxide from the atmosphere, or water from the soil. Animals and fungi could feed upon other living things for nourishment. Doing so required plants and animals to transform matter. "We ought, therefore," concluded Weiss, "to look upon assimilation rather than growth as the attribute of life."[33] In the context of heredity, assimilation was very similar to hybridization. The young professor would address this equally fundamental concept only a few years later.

Weiss came from a background where assimilation was a necessary part of life. His father was a German émigré who had lived in the Netherlands. As a liberal, he had refused calls to return to the country of his birth for conscription when the army was being used to crush the 1848 democratic uprisings. His mother, of Huguenot ancestry, encouraged the young Weiss to take an interest in natural history. Although Weiss was born in Britain, he received much of his education in Germany and Switzerland.[34] He went on to study zoology and botany at University College London, before taking up posts across Europe to study such diverse topics as marine organisms and rubber plants. At the age of twenty-six he was appointed professor of botany at Owens College. Manchester and its surrounds offered many opportunities for an academic botanist. Weiss was a regular figure in local societies and an active visitor to the coal deposits of Lancashire and Yorkshire. From 1902 to 1911 his primary scientific interest was locating and examining the remains of fossilized plants and reconstructing their morphology and their relationship with ancient environments.[35]

As in agriculture, the rediscovery of Gregor Mendel's laws did not change biology overnight. Their influence on the fossil-hunting Weiss also came slowly. However, in 1911 Weiss was elected president of the Manchester Literary and Philosophical Society and chose Mendel as the topic of his inaugural address. He told his audience that he was particularly intrigued by the famous "three to one" ratio that Mendel had observed among his population of smooth and wrinkled peas. Weiss expressed his amazement that "the numerous investigators who have continued Mendel's experiments in a great variety of plants and animals have in almost every case obtained the same numerical ratio."[36] Applying Mendelism to farming could also produce remarkable results. Weiss noted that leading varieties of British wheat, a vital wartime commodity, were vulnerable to fungal infections and did not possess as much gluten as their North American counterparts. These shortcomings had been overcome by Rowland Biffen, chair of agricultural botany at the University of Cambridge. With the aid of Mendelian genetics, Biffen had united immunity and high levels of glucose to create new varieties of wheat that thrived in the wet British climate.[37] Taking heart from this example, Weiss concluded his address on a grandiose note: "A general understanding, indeed, of the laws of inheritance, which seem to be so remarkably uniform in the vegetable and animal kingdoms, is not only essential for the advance of the biological sciences, but is a basis for the economic and social progress of mankind."[38]

For many of his contemporaries in the biological sciences, the wholehearted acceptance of Mendel came with a strong aversion to any phenomenon that might suggest an alternative vision of heredity. But this was not the case for Weiss. In 1916, at the height of the First World War, he appeared before the Manchester Microscopical Society once more. Weiss had been elected president of the society and he chose the controversial topic of graft hybridization for his latest presidential address. At this point, we might also consider why Weiss would choose to engage with a highly contentious question at a cosmopolitan scientific society, rather than in a more formal academic environment. Philosophical societies, particularly those with an inclination toward natural history, had been a part of the British scientific scene for centuries. Many of these societies had a civic purpose, seeking to educate the urban populace through museums and lectures while providing their members with an uplifting and moral hobby. Microscopical societies, however, were less invested in civic education and more in facilitating discussion between enthusiastic microscopists.[39] The Manchester Microscopical Society had close links with the University of Manchester. Weiss himself had been introduced to the society in 1893 by his colleague Professor Milnes Marshall. An experienced alpine climber, Marshall met an untimely end in the Lake District. His demise, falling in an attempt to take a photograph, had a distinctly modern flavor.[40]

Weiss's own account of the Manchester Microscopical Society, which he produced in 1930, gives a sense of an organization steeped in history and intrigued by both living and fossilized plants. Some hints of at least an inclination toward Lamarckism were also present. One example was Thomas Hick, member of the society and headmaster of the Lancastrian School in Leeds, who had used microscopy to argue for the existence of protoplasmic continuity in algae. On the strength of this research, he was appointed as a lecturer in botany at Owens College in 1885. He was also interested in plant fossils and assisted Weiss upon the latter's arrival at the college.[41] Despite its exclusively scientific focus, the Manchester Microscopical Society was not immune to the political controversies of the early twentieth century. On one occasion, the admission of women to the society proved such a heated topic that Weiss was called in to act as an impartial arbitrator. He recalled:

> It was when Mrs. Leo Grindon was proposed for membership, and the Society was sharply divided into those anxious to maintain a purely masculine membership, and those, of shall we call them advanced views, or shall we say Bolshevik tendencies who wished or were prepared to admit women. At any rate, there was an upheaval and some feared a revolution. I presided at the second somewhat heated discussion. Mrs. Grindon was black-balled for membership, but was invited to attend the meetings, if she liked to do so. This was a display of the typical British love of compromise. A little later Miss Annie Dixon became a full and active member, and since then other ladies have joined. I presume that now that political equality has been granted to both sexes, no properly qualified woman would be denied membership.[42]

In this instance, Weiss was no radical. However, he was certainly no counterrevolutionary either. His neutrality in this controversy, combined with the society's interest in plant physiology and structure, most likely made for a friendly setting in which Weiss could safely argue in favor of the existence of graft hybrids. His standing as the society's president probably did not hurt. Weiss began his 1916 address on graft hybrids by noting the antiquity of grafting and acknowledging that botanists were now "inclined to be more skeptical of the existence of graft hybrids than were the botanists of earlier days."[43] Despite this newfound skepticism, however, Weiss saw a wealth of historical evidence and modern experimentation that supported graft hybridization. He began with a crowd-pleasing reference to Charles Darwin, the famed English naturalist. Darwin, explained Weiss, had been a great believer in graft hybridization, conducting numerous experiments by crossing potato varieties. Weiss also supported Darwin's interest in the infamous *Cytisus adami*, the nineteenth-century graft hybrid later outed as a chimera. Disputing conventional wisdom, Weiss argued that the status of the plant remained unresolved. He noted that nobody had ever been able to produce a seed hybrid of it, suggesting instead that cell fusion might have occurred around the graft junction. Weiss ended his discussion of this controversy by concluding, "It seemed therefore as if our efforts to discover the true nature of *Cytisus Adami* must remain fruitless."[44]

Another familiar individual who influenced Weiss was the German botanist Hans Winkler. Members of the Manchester Microscopical Society were consequently treated to a perspective of the Winkler-Baur controversy that many European botanists would not have recognized. Weiss was well informed of the intricacies of the controversy, describing how Winkler had combined tomato and common nightshade to create multiple hybrids. He also described Baur's rebuttal, which he characterized as largely based upon the "skin deep" observation that Winkler's hybrids did not show any mixing of tissue on their exterior. In fact, argued Weiss, it was entirely possible that different layers of cells in a grafted plant could intermingle. Here he referenced the Cambridge botanist Margaret Hume, whom he described as demonstrating that "some internal inter-action might to some extent take place between the living substance of the two species."[45] In fact, Hume had showed no indication that the unity between layers of cells in plant chimeras could result in the exchange of hereditary material. Weiss concluded his account of the Germanic controversy with reference to the famed and complex chromosome count of Winkler's *Solanum darwinianum*.[46]

On a global scale, biological and agricultural experimentation offered the tantalizing possibility of graft hybridization. Weiss was aware of the near-mythical status of the French botanist Lucien Louis Daniel, who had declared that hybrid plants produced by the crossing of seeds were essentially identical to hybrid plants produced by grafting. Daniel had argued that the formation of distinct layers, or sections, of different tissues in plant chimeras could be seen to occur in plants that had been conventionally hybridized, not grafted. To Weiss, this observation indicated that the distinction drawn between chimeras and graft hybrids in Germany was artificial and did not reflect the complexities of biology.[47] To gain more evidence in support of the hypothesis that conventional breeding could result in tissue segregation, Weiss looked further afield, turning the attention of his Manchester audience to Japan and the silkworm.

At the Zoological Institute of Tokyo Imperial University's College of Agriculture, silkworms were an organism of considerable interest. The production of silk and silkworm breeding were undergoing a transformation as Japan opened to global markets in the early twentieth century. Kametaro Toyama, who completed his doctoral dissertation on hybridization in insects, had applied Mendelian genetics to silkworms in 1905. Some aspects of Toyama's claims attracted immediate controversy, with the American geneticist and breeder Vernon Kellogg arguing that the colors of silkworm cocoons did not reflect Mendelian principles. Kellogg blamed this irregularity on the long history of silkworm cultivation, pointing out that cultivators had traditionally focused on selecting aesthetically pleasing cocoons in the belief that these would produce better silk. Silkworm larvae, on the other hand, had been largely ignored by breeders and so "naturally" adhered to Mendelian genetics.[48] Unlike Kellogg, Weiss was no critic of Toyama. He instead zeroed in on a particular part of the latter's research that would aid his case for graft hybridization. When Toyama had crossed a white Japanese silkworm with a striped French variety, the resulting caterpillars were striped on one half of their bodies but plain on the other. This, declared Weiss, supported his theory that characters did not necessarily "blend" in a hybrid. A plant or animal with the outward appearance of a chimera could therefore be a sexual or graft hybrid.[49]

A small network of correspondents soon emerged around Weiss's interest in graft hybrids, which directed Weiss toward promising plants. Leonard Alfred Boodle, assistant keeper of the Jodrell Laboratory at the Royal Botanic Gardens, Kew, met Weiss at the gardens around the time of the latter's 1916 address to the Manchester Microscopical Society. "When you were here," wrote Boodle to Weiss in September 1916, "you said that you might be coming to Jodrell Lab. later to examine some of the graft hybrids."[50] The graft hybrids in question may have been specimens of the "Bronvaux hybrid," which Weiss would later describe to the Manchester Literary and Philosophical Society. Only a month after his letter from Boodle, Weiss was exchanging notes with Nordal Wille, botanist and manager of the botanic

FIGURE 7. *Cratægomespilus dardari*. In the mid-1920s, German-British botanist Frederick Weiss examined several plants which had been made by grafting the branch of a medlar tree to the stock of a hawthorn tree. German botanist Erwin Baur had described these medlar-hawthorn crosses as chimeras. Weiss disagreed and declared that they were graft hybrids. Lucien Louis Daniel, *Les mystères de l'hérédité symbiotique: Points névralgiques scientifiques, pensées, théories et faits biologiques* (Rennes: Roger Gobled, 1940), plate XI.

garden in Christiania (now Oslo), Norway. Weiss soon received an interesting delivery. "I have sent you as a post-parcel," wrote Wille in October 1916, "a twig with leaves of the supposed graft hybrid between Crataegus oxyacantha [a species of hawthorn] and Pyrus communis [the common pear] from Torp in Norge [Norway]." The remote specimen, which rarely fruited or flowered, had taken him some time to procure. "I now think," reported Wille, "that it is doubtful whether it really is a graft hybrid." The Norwegian botanist did not expand on why he now thought that the "Torp tree" was not a graft hybrid. He had described it as such in 1908, explaining how the hybrid had been "discovered by an apothecary late in the eighties" between the towns of Fredrikstad and Sarpsborg.[51]

Flirting with alternatives to Mendelian genetics did not damage the career of Weiss, who served as temporary vice-chancellor of Manchester University during the early years of the First World War. In 1924 Weiss returned to the Manchester Literary and Philosophical Society. The Royal Botanic Gardens, Kew had since sent him a sample of the Bronvaux hybrid, *Crataegomespilus*, which had once fascinated the American geneticist George Harrison Shull. Hawthorn made up the interior of the plant, with its exterior tissue derived from a medlar tree. Baur, the steadfast German opponent of graft hybrids, had previously examined the plant and dismissed it as a chimera. Weiss, however, was not convinced by Baur. Under his microscope, the cells of the Bronvaux hybrid were very different from either of its parent plants. Weiss found the same differences in another variety of *Crataegomespilus*, which he had obtained directly from Daniel (fig. 7). "The features revealed bear out the conclusions of Daniel," Weiss concluded, "that in graft-hybrids we have a similarity with the phenomena noted in the case of seed-hybrids."[52] By this, he was referring to the tendency of some sexual hybrids to segregate into different tissues, the same phenomenon described by Toyama in silkworms. By 1926 he could articulate the purpose of graft hybridization research and its relation to genetics:

> Since the re-discovery of Mendel's laws of inheritance, innumerable exper-
> iments have been carried out on plants and animals with a view to testing
> the universality of the phenomenon of dominance and segregation which
> Mendel had demonstrated by his experiments on peas. The outcome of this
> activity has been to confirm in the main Mendel's work. But it is natural
> that irregularities and exceptions have been discovered and the explanation
> of these has resulted in a modification of the simplicity of the theory as

originally set forth. Attention, indeed, in all such cases tends to become concentrated upon the phenomena which do not fit into any generally adopted scheme, and no doubt much can be learnt from such exceptions.[53]

By the late 1920s exceptions to the usual Mendelian rules were apparently multiplying. The John Innes Horticultural Institution, former home of the recently deceased William Bateson, was no exception. A student there, Reginald J. Chittenden, discussed the relationship between genetics and such phenomena as plant chimeras in 1926. "The endeavour to force a Mendelian explanation on the reluctant facts of mosaicism [a kind of chimerism]," he wrote, "has necessitated the formation of many subsidiary hypotheses, all involving alterations in, or modifications of the generally accepted and well substantiated views on the mechanism of heredity."[54] These modifications included such concepts as the *labile gene*, a gene with an extremely rapid mutation rate. Paying tribute to the late Bateson, Chittenden admitted that "in some cases views have been expressed here of which he would disapprove. In the main however, the interpretations adopted are those already publicly expressed by him."[55] Naturally, Weiss embraced this suggestion of a more permissive approach to genetics.

In 1930 Weiss took his views on graft hybrids from the friendly societies of Manchester to the cutthroat world of academic publishing. A lengthy piece in *Biological Reviews* bore the classical Weissian hallmarks of careful documentation and conservative conclusions. He also addressed animal graft hybrids for the first time. "Less has been done in the animal kingdom in this direction," he noted, "as grafting is not a universal practice, as it has been from time immemorial in horticulture." Amphibians and aquatic animals, including salamanders and hydras, had been the basis of grafting experiments. For now, noted Weiss, these experiments had succeeded only in producing "chimaeric organisms, sometimes of sectorial character."[56] He did not reference Guthrie or Kammerer, either out of ignorance of their work or in an attempt to avoid association with "disreputable" research. Weiss did include a succinct summary of why graft hybridization had become such an issue in the twentieth century:

> The renewed interest which was awakened in the study of graft hybrids
> early in the present century, no doubt in association with the revival in
> that of genetics brought about by the "rediscovery" of Mendel's work, has
> led to a considerable development of our views as to their nature. Cytology,

morphology and genetics have contributed towards the progress. Winkler's experimental work in the production of graft hybrids in *Solanum* and Baur's experimental work with *Pelargonium* have been particularly helpful, and the suggestion that both graft hybrids and white-margined leaves should be regarded as periclinal chimaeras, while it has produced plentiful criticism, has also suggested new lines of investigation, and has proved to be as fruitful as any fundamentally new conception usually is, whether it ultimately stands the test of time or not.[57]

Genetics, then, had spawned its own antithesis in the form of a new wave of graft hybrids. Weiss was clearly not convinced that the chimera hypothesis could explain away every instance of observed hybridization through grafting. "There is no doubt that the problem of graft hybrids is by no means completely solved," he wrote, "though we have a better conception now of the possibilities of these curious formations." To pursue this problem, Weiss came up with his own recipe for the creation of future hybrids. He suggested that botanists conduct grafting experiments with dioecious plants, where male and female organs appear in separate individuals instead of on the same plant. This separation, he speculated, "might presume a greater tendency of fusion of the vegetative tissues of a male and a female plant."[58] Weiss's effort to expand graft hybridization beyond the world of Manchester societies elicited little reaction from his academic peers. Why was Weiss not condemned when Kammerer, for instance, was? Several factors played into Weiss's hands. For one, he was already a certified Mendelian with an interest in unusual aspects of heredity that genetics struggled to explain. From this perspective, he was wrestling with the same questions that Bateson had grappled with at the John Innes Horticultural Institution. Unlike Kammerer, Weiss diplomatically depicted graft hybridization as a problem for biologists to solve, rather than as a counter to genetics. His considerable academic standing also helped his cause.

Until his retirement from academia, graft hybrids remained on Weiss's mind. At a special two-part lecture in 1940, delivered to the Royal Horticultural Society, he acknowledged that most historical examples of graft hybridization could be explained through the chimera hypothesis. Even a pear-quince tree produced by Daniel did not escape. Weiss explained that its unique leaves, with a smooth edge at the base and a serrated edge at the tip, could be explained by the tendency of the base of leaves to contain material from the core of the plant (in this case, quince) while the tips contained material from the outer layers (which in this case resembled pear).[59] "The occurrence of a real graft hybrid," mused Weiss, "in which actual cells of stock and scion have fused together like the fertilization of an egg cell by a pollen grain is still very problematical."[60] The content of his lecture suggested that the members of the John Innes Horticultural Institution had been hard at work convincing Weiss that chimeras could explain almost all horticultural phenomena related to grafting. Chimeras were both more common and more varied than Weiss had previously realized. While "we have gained considerable insight into the general character of chimaeras," he concluded, "whether produced by grafting or by other means, there are still many problems to be solved before our knowledge of these interesting productions is complete."[61]

What insights can we glean from Weiss and his support of graft hybridization over a twenty-five-year period? His career reveals that it was possible to walk the tightrope between Mendelian genetics and graft hybridization for much of the interwar era. This tightrope, however, unraveled over time. Its existence was only possible thanks to three factors unique to the 1920s and 1930s. Firstly, Mendelian genetics, even with the addition of Morgan's chromosome theory, became briefly more inclusive. At the John Innes Horticultural Institution—hardly a bastion of scientific unorthodoxy—chimeras became research objects, while the fixed nature of the gene itself was questioned. Secondly, graft hybridizers such as Weiss thrived thanks to ongoing ambiguity over results of transplantation and grafting experiments, both historical and contemporary. He could, for instance, interpret the "Mendelian" experiments of Toyama on silkworms as providing evidence in favor of graft hybrids. Finally, metropolitan societies provided a supportive setting in which to explore ideas not yet ready for academic science. Hence Weiss turned to the civic societies of Manchester to explore the big questions, from the nature of life itself to whether natural barriers to the transformation of life could be breached through graft hybridization.

Ghost of a Hypothesis

From 1916 to 1948 William Neilson Jones led the Botany Department at Bedford College in London. The college had been founded in 1849 by Elizabeth Reid, a Unitarian who wished to provide women with an education for the benefit of society. In the early years of the twentieth century, botany and geology shared scarce space and equipment at Bedford. Both were taught by Catherine Raisin, a noted geologist and advocate of equal education for women.[62] By 1900 the college became part of the wider University of London and was attracting more and more talented students and lecturers. Among the latter was William Neilson Jones. In 1910 an interest in *Calluna vulgaris* (common heather) brought Neilson Jones into contact with the botanist and mycologist Mabel Rayner, whom he married two years later.[63] At the outbreak of the First World War, Bedford College received an endowment from politician and businessman Hildred Carlile. Neilson Jones became head of the now-separate Botany Department and received the title of Hildred Carlile Professor. Initially, he would use his position to enhance his status as a botanical educator, pushing a curriculum that included Mendelian genetics.

A textbook that Neilson Jones coauthored with Rayner, published in 1920, would give its reader the impression that Mendel reigned supreme in the mind of Neilson Jones. The two authors hoped that their *Textbook of Plant Biology* would "enable the student or intelligent layman to acquire an understanding of the relation of plant life to general biological knowledge," by presenting modern science in "simple language."[64] Grafting appeared only briefly, in a comment on propagation methods. Mendelism, on the other hand, received almost an entire chapter. "The study of heredity, variation and everything relating to this aspect of living organisms," explained Neilson Jones and Rayner, "is known as the science of *Genetics*, —a recent branch of Natural Science which, apart from its absorbing theoretical interest, has already yielded results of great practical value."[65] For the authors of educational textbooks, the focus of Mendelian genetics upon experiments and practical results was quite attractive. Introducing Mendel's experiments on peas, Neilson Jones and Rayner remarked, "It is convenient to commence our study of the problems of inheritance in plants with a description of an actual breeding experiment."[66] They went on to praise Mendel as a scientific pioneer and to remark on the close similarities between the behavior of chromosomes and Mendelian "units," or "factors." This close relationship strongly suggested that chromosomes formed the basis of heredity, which in turn could be predicted and controlled through the application of genetics.

There was one major caveat to the power of Mendelian genetics. Although "the behaviour of a very large number of characters has been investigated by the Mendelian method of studying heredity," wrote Neilson Jones and Rayner, "in the majority of cases the characters studied are such as distinguish races or varieties of plants from one another."[67] A mountain of evidence supported Mendelian genetics when it came to the inheritance of relatively minor characteristics, such as the color of peas or horn length of cattle. There was less evidence that Mendelian genetics held true when different species were crossed. The natural limitations to sexual hybridization were part of the problem. "When any but fairly close relatives are crossed," noted Neilson Jones and Rayner, "either no offspring is obtained or it is impossible to carry the experiments further owing to the offspring being sterile."[68] Crossing different species was still of economic interest, as their hybrid offspring would be larger and more robust than the parent types. Hybrids of this sort, however, were hard to produce and unlikely to be amenable to further experiments. Enter graft hybridization, which offered its practitioners the opportunity to overcome natural limitations on species crosses and produce new plants of high economic value.

Given the potential of graft hybridization to contribute to heredity theory and agriculture, the controversy over the existence of graft hybrids naturally formed part of the botany curriculum at Bedford College. In the published version of his lecture course, released in 1934 as *Plant Chimaeras and Graft Hybrids*, Neilson Jones again announced his intention to reach a broad audience. "Much of the material in the present volume," he reported, "formed the subject matter of a course of inter-collegiate lectures to advanced students of botany in the University of London." However, he reported that "it is believed that the facts are sufficiently arresting and many of the examples of such general occurrence as to appeal to a wider public, including those intelligently interested in plants although not themselves botanical specialists." Neilson Jones began somewhat incongruously, adopting a liberal interpretation of Hume's work to suggest that grafted plants could form cellular connections.[69] In spite of his efforts to reach a larger audience, he cautioned against simplistic grafting experiments and warned of the dangers of jumping to conclusions:

> How necessary it is to exercise caution in the interpretation of an experiment is shown by the following. If a branch of a green-leaved variety of some plant is grafted on a red-leaved stock, it sometimes produces red leaves. The hasty deduction might be made that the faculty for producing red pigment had been transmitted from stock to scion. Such a conclusion would be unwarranted, however, for further experiment would show that the production of red leaves is equally evident when the branch is grafted on a green-leaved stock! The development of red pigment in leaves is greatly influenced by conditions of water supply, being favoured by water scarcity.[70]

From the outset, Neilson Jones acknowledged that chimeras existed and were composite organisms that could emerge from a graft junction. He cautiously described graft hybridization as a "hypothesis" and declared that it could occur only following cell fusion of scion and stock "in a manner analogous to that by which fusion between gametes or sex-cells gives rise to a seed hybrid." To distinguish these genuine graft hybrids from their controversial forebears, he labeled such a plant a *burdo*.[71]

In forming this description of graft hybridization, Neilson Jones drew heavily upon the work of Daniel. The French botanist had argued that variations could occur in grafted plants that had not been present in either stock or scion. These variations could then be inherited. The chances of obtaining a new plant could be improved by following Daniel's famous grafting technique—*la greffe mixte*—which involved encouraging fresh plant shoots just below the original grafting site to develop. Neilson Jones also noted that Daniel's views "are not in accord with those noted as having met with general acceptance," with most botanists explaining away new variations in grafted plants as the result of "disturbed nutrition."[72] Neilson Jones was not deterred, arguing that linking cell fusion and graft hybridization had theoretical advantages: "There are two things to be said in favour of a graft hybrid conceived as a special form of plant—or burdo—resulting from vegetative nuclear fusions between scion and stock: in the first place the conception is perfectly clear and definite; secondly, as will be argued later, there is nothing in such a conception that is impossible on theoretical grounds."[73]

Neilson Jones saw his description of graft hybridization as a clear contender against the rival "chimera hypothesis." When he gave his lectures at Bedford College during the 1920s and 1930s, however, he was not sure that graft hybrids even existed. While there were no grounds to reject the theoretical existence of graft hybrids formed by cell fusion, there was not strong enough evidence to accept their existence either. The difficulty in settling the graft hybrid question lay in demonstrating whether cell fusion had occurred between genetically distinct cells. Such events were rare and difficult to spot. Neilson Jones argued that counting chromosomes could offer only indirect evidence of graft hybridization. The appearance of many chromosomes in a plant cell was not incontrovertible proof that these chromosomes had arisen from fusion with the cell of another plant. "The existence of graft hybrids therefore may be regarded as theoretically possible and the conception may be useful as a working hypothesis," he told his students, while "at the same time, although the reality of their existence may not be definitely disproved by any known facts, it yet lacks positive support from unimpeachable evidence."[74]

By applying his chromosome skepticism to the famed case of Winkler's *Solanum darwinianum*, Neilson Jones concluded that a neutral observer could not rule whether the plant had been a graft hybrid or not. In his retelling of the controversy, Neilson Jones described how the plant was found to possess forty-eight chromosomes, an unusual number found in neither its parental tomato nor nightshade. Still, forty-eight was only half the sum of chromosomes found in tomato or nightshade, an inconvenient fact attributed by Winkler to the plant's chromosomes being "halved by some auto-regulatory process."[75] Neilson Jones was not convinced. He knew that mutations that doubled or quadrupled chromosomes frequently appeared in injured plants. As forty-eight chromosomes was twice the number found in a tomato, it was possible that *Solanum darwinianum* held mutant tomato cells, rather than hybrid ones. Winkler was aware of this fact and had continued to insist that the *Solanum darwinianum* displayed the characteristics of a true hybrid.[76] As ever, the truth about the plant remained hidden, although Neilson Jones erred toward it being a mutant chimera. He also dismissed other past examples of graft hybrids, including *Cytisus adami* and the Florentine Bizzaria, as chimeras.

The dismissal of the graft hybrids was unlikely to have been a painful decision for Neilson Jones. Like Weiss and other botanists of the 1920s and 1930s, he did not agonize over the distancing of genetics from heritable changes directly induced by the environment. On a more practical level, too, chimeras could also be of great economic importance to horticulturalists. Neilson Jones gave the example of an incident in 1928, when a sudden European frost had killed off tomatoes. The only survivors were chimeras, tomato plants that had been grafted onto a perennial scrub called *Solanum memphiticum*. The scrub had formed a protective "skin" around the tomato plants, saving them from the cold. Such observations suggested that resilience could be imposed onto vulnerable crop plants through grafting, albeit via physical, rather than genetic, means. Neilson Jones reported that one horticulturalist had sought to apply this insight to potatoes, which were vulnerable to infection by the *Phytophthora* genus of molds, one member of which was responsible for potato blight. The ultimately unsuccessful plan was to graft a tomato and a potato plant, with the skin of the tomato effectively granting immunity to the potato inside.[77] In a way, chimeras could potentially confer many of the benefits to horticulture once promised by graft hybrids.

Upon its release in 1934, *Plant Chimaeras and Graft Hybrids* received almost universal praise. The journal *Science Progress* reported that "Prof. Neilson Jones has here given us an admirable and very readable account of these interesting plants, which have as it were a dual personality."[78] The *Irish Naturalists' Journal* went further, not only praising the "very readable exposition" but exclaiming that "plants with tomato cores and fruits but a nightshade epidermis promise almost as much excitement as the best detective tales." In fact, "when one adds a heated controversy," continued the reviewer, "between Baur with his chimaera hypothesis and Winkler with his vegetative fusions or true hybrid hypothesis, the interest should be obvious."[79] Someone who probably did find the graft hybrid debate as exciting as a detective story was Weiss. In December 1934, Weiss wrote his own review for the *New Phytologist*. In it, he noted that Neilson Jones had effectively sided with Baur and classed most historical graft hybrids as chimeras. As Weiss himself had independently adopted this view, he did not protest. He also agreed with Neilson Jones that the *Solanum darwinianum* was likely a chimera, containing a core of tomato cells with extra sets of chromosomes. On the likelihood of a true graft hybrid ever coming to light, however, Weiss was more upbeat. He repeated his suggestion that botanists could conduct grafting experiments with plants where the male and female organs were in separate individuals. "Perhaps if Winkler's experimental methods were to be undertaken with two dioecious plants," he wrote, "one male and the other female, and possessing the same number of chromosomes, a fusion of vegetative cells might be obtained."[80]

The economic value and scientific interest of chimeras, not graft hybrids, would come to dominate Neilson Jones's thoughts. In a 1937 article he expanded the list of practical uses for plant chimeras. The plants, he stated, had been valued only as ornaments or curiosities. Yet the ability of chimeras to form a core of one plant surrounded by the skin of another gave rise to the possibility that a chimera could resist diseases usually fatal to an individual plant. In Russia, popular types of potato, such as the Golden Wonder, were found to be chimeras, with a core of one variety and the skin of another. Although Neilson Jones did not know whether the skin of the Golden Wonder acted as a "resistant barrier for more susceptible core tissue beneath," he was aware that the different plants that composed some ornamental chimeras had maintained their characteristic immunity following grafting.[81] On the downside, it was also possible that a plant might pass on its susceptibility to certain diseases. When some of Winkler's tomato-nightshade chimeras had been exposed to a fungal disease that attacked tomatoes, some had succumbed while others had proven immune.[82] "Such results as these," concluded Neilson Jones, "point to mutual reactions of the chimaeral components that may affect their natural immunity." Such reactions, however, were not evidence of graft hybridization. In addition, their unpredictable outcomes led "to the need for caution in attempting to forecast precisely the degree of resistance of a chimaera to pathogenic attack."[83]

Following the Second World War, levels of tolerance for graft hybridizers within the Western scientific community fell rapidly. As we will see, support for graft hybridization within the Soviet Union bitterly divided biologists. Neilson Jones held something of a sympathetic attitude to the strange agronomy that emanated from the Soviet Union. In his 1947 book *The Growing Plant*, he decided not to be overtly critical of leading Soviet agronomist Trofim Lysenko. On Lysenko's ridiculed assertion, for instance, that plants passed through defined phases in their life cycle, Neilson Jones chose to be neutral rather than dismissive. "Recent work," he noted, "has shown that the reactions of plants to their surroundings are far more complex than was realized when Lysenko's views were first propounded."[84] Arguing that international botanical controversies were soon left behind by the march of science was certainly easier than dealing with profound ideological or scientific disagreement. Independently of the Lysenko affair, Neilson Jones had largely abandoned the graft hybrid hypothesis. His 1937 article on chimeras did not mention graft hybrids, and he chose not to speculate about them in *The Growing Plant*, instead sticking to known facts about chimeras.

When a second edition of Neilson Jones's *Plant Chimaeras and Graft Hybrids* was released in 1969, graft hybrids had been stripped from the title. This change likely reflected a growing Cold War polarization over their existence. It also aligned with Neilson Jones's own beliefs, which had begun moving away from the existence of graft hybrids three decades prior. If its author hoped, however, that a superficial title change could save him from Cold War controversy, he was in for a rude awakening. Harold W. Howard, an agricultural botanist at the Plant Breeding Institute in Cambridge, was scathing. "Although graft hybrids has been omitted from the title of the second edition and although, as the author states, 'the burdo is at present an entirely hypothetical structure,'" noted Howard, "the ghost of the graft hybrid hypothesis haunts much of this book."[85] Even more brutally, Howard recommended that the book would benefit from a complete rewrite and directed prospective students of botany to other sources. Other reviewers

praised Neilson Jones for abandoning the graft hybrid. The plant morphol-
ogist Frederick Albert Lionel Clowes observed that, despite its age, the first
edition of the book was still consulted by students for information about
chimeras. "In those days," he wrote, referring to the 1930s, "graft hybrids
or burdos resulting from the supposed fusion of somatic nuclei were still a
probability. The burdo theory dies hard." Although Neilson Jones was still
"less damning than most geneticists are," he had at least "clearly come out
wholly in favour of the chimera hypothesis to explain the particular plants
sometimes obtained by grafting."[86]

After more than thirty years in botanical education, Neilson Jones had
still failed to shed the disreputable ghost of the graft hybrid. He had begun
with a skeptical attitude toward their existence, describing the experiments
of Daniel while acknowledging that "many of his views would not be met
with general acceptance among botanists."[87] On the existence of animal
graft hybrids, too, Neilson Jones reserved his judgment. Although animal
transplantation and immunology had produced some interesting results, he
concluded, "caution is required before transferring all the machinery of animal serological reactions to the plant kingdom."[88] The graft hybrid was only
ever an intriguing hypothesis. It could be safely discussed as a hypothetical,
in the same way that Galileo debated Copernican theory. Throughout his
life, Neilson Jones showed no tendency to take graft hybridization more
seriously. Like Weiss, in fact, he moved further and further away from the
graft hybrid as the twentieth century progressed. Part of this movement
reflected advances in botany, as scientists outed grafted plants as chimeras
and gleaned new insights into the workings of mutations and chromosomes.
Neilson Jones, unlike many of his contemporaries, also considered chimeras
to be a subject worthy of practical study. They could essentially achieve
many of the same results of graft hybridization, including the exchange of
environmental resilience and disease resistance between species, without any
theoretical or ideological baggage.

A Persistent Idea

So far, the story of graft hybridization has had a distinctly rural flavor.[89]
The key players in the American controversy over graft hybrid poultry were
all the products of family farms: agriculturalists fighting over the future
of agriculture. The clash between Hans Winkler and Erwin Baur played
out as much in botanic gardens and horticultural stations as it did in the
laboratory. As the end of the first half of the twentieth century approached,

however, defenders of graft hybridization were most likely to be found in the academic institutions of urban spaces. Their support for a controversial idea was articulated in zoological laboratories, the meeting spaces of scientific societies, and university lecture halls. Urban spaces were not just environments where science happened. The "practices and characteristics of the urban environment" also shaped how scientists chose, organized, and validated their research.[90] Likewise, our three graft hybridizers adopted different scientific and career strategies that reflected their environments. The opportunities and dangers offered by interwar Vienna would make and break Paul Kammerer. The industrial landscape and heritage of Manchester gave Frederick Ernest Weiss the space to dabble in the practical and philosophical aspects of biology. In London, amid a densely packed literary and educational environment, William Neilson Jones placed strict limits on acceptable standards of scientific evidence and who was best placed to assess it.

Although the wealth and patronage of Vienna gave Kammerer the opportunity to conduct his graft hybridization experiments, the cultural and scientific landscape of the city had significant pitfalls. The scientists of the Vivarium were "bound by shared experiences," from education to exclusion, which "established a degree of homogeneity in their daily lives."[91] One result of this homogeneity at the Vivarium was the hunt for patterns or analogies in nature.[92] Kammerer naturally joined this hunt, publishing a popular book on coincidences titled *Das Gesetz der Serie* (The law of series). Taking an institutional obsession into the wider world had consequences. To a committee assembled in 1918 to consider Kammerer's application for a professorship, the book "smacked of mysticism and pseudo-science" and likely cost Kammerer the academic post.[93] During the later controversy over the midwife toad, the dark side of Vienna emerged in the form of grotesque antisemitism. A group of antisemitic professors who operated out of Vienna University may even have had a hand in the midwife toad affair. The scientific and cultural landscape of interwar Vienna birthed Kammerer's grafting experiments, but also repressed his reputation and legacy.

The once rootless Weiss found his urban home in Manchester, an industrial city with a deep-seated scientific culture. The legacy of the Industrial Revolution in the region left Weiss with easy access to fossilized plants in coal deposits. From these fossils, Weiss gained an appreciation of how closely the environment and its botanical inhabitants were tied together. The rich civic scene of Manchester, with its scientific and philosophical societies, gave Weiss the opportunity to explore aspects of biology with economic or social implications. For city authorities, scientific meetings represented progressive values and technical expertise.[94] For Weiss, such meetings enhanced his social status and allowed him to test new ideas. He gave Mendelian genetics a glowing review at the Manchester Literary and Philosophical Society in 1911, even as he was also interested in exceptions to genetic laws. By the mid-1920s Weiss was able to articulate what he believed about graft hybrids and why they were still worthy of scientific interest. The scientific societies of Manchester gave him a setting in which he was able to build his arguments and present them to an audience, all before they appeared in international publications.

Finally, we come to Neilson Jones. As a member of a progressive London-based community of scientists and educators, Neilson Jones articulated the practical benefits of biology in his Bedford College lectures. Mendelian genetics featured prominently. Yet Neilson Jones quickly cottoned on to the fact that sexual crosses had strict limitations as a hybridizing tool. Graft hybrids and chimeras offered a way to overcome these limits and more freely exchange beneficial traits between species. Acutely aware of the intense controversy over graft hybridization, Neilson Jones attempted to demarcate what evidence—and by implication, whose expertise—would be included in the debate. Horticulturalists and gardeners who believed in the influence of stock on scion thanks to the changing color of leaves were wrong to rush to conclusions. Further experiments, ideally conducted by a university-trained botanist, would demonstrate that such observations were irrelevant to the graft hybrid debate. Of our three interwar graft hybridizers, Neilson Jones was the only one to remain an active participant in the scientific community into the mid-twentieth century. He was, in one sense, unfortunate. As the Cold War progressed, support or denial of graft hybridization became a matter of keen ideological contention.

FIGURE 8. Potato grafted onto tomato. At the meeting of the V. I. Lenin All-Union Academy of Agricultural Sciences in 1948, Trofim Lysenko displayed wax models of what he claimed were potato-tomato hybrids created by grafting. Although such hybrids were not new, Lysenko used their existence to attack Mendelian genetics. Lucien Louis Daniel, "Sur la réussite, le développement, la durée et la production des greffes (suite)," *Revue bretonne de botanique pure et appliquée* 6, nos. 2–3 (1911), plate 2.

4

Beyond Lysenko

IN WHAT THE SCIENCE WRITER STEPHEN JAY GOULD WOULD CALL "the most chilling passage in all the literature of twentieth-century science," Trofim Lysenko addressed the V. I. Lenin All-Union Academy of Agricultural Sciences on July 31, 1948, with a speech preapproved by the Central Committee of the Bolshevik Party.[1] To thunderous applause, the peasant-farmer-turned-leading-agronomist unveiled his vision of a new Soviet biology: the rejection of Mendelian genetics; denial that the chromosome was the seat of heredity; promotion of the inheritance of acquired characters; and supposedly new agricultural techniques, including the vernalization of wheat. Lysenko also used his address, which had been edited by Joseph Stalin himself, as an opportunity to attack his opponents. The outspoken and rebellious botanist P. M. Zhukovsky was first in the firing line. "As becomes a Mendelist-Morganist," Lysenko exclaimed, "[Zhukovsky] cannot conceive transmission of hereditary properties without transmission of chromosomes. . . . He therefore does not think it possible to obtain plant hybrids by means of grafting."[2] At this point, Lysenko gestured toward the back of the hall, where potato-tomato hybrids obtained through grafting stood (fig. 8).[3] The graft hybrid had taken on a new political life. Just as traditional hybrids had once been used as a practical demonstration of the utility of Mendelian genetics, now graft hybrid plants were held up as living examples of the truth of Soviet biology, including the inheritance of acquired characters.

The chilling Academy of Agricultural Sciences meeting was the height of the "Lysenko affair," now a model example of the consequences of political ideology triumphing over science.[4] Yet, as we have seen, there was nothing original in Lysenko's proposal that new species could be created through graft hybridization. A decade or so earlier, such a proposition would have found support from several prominent biologists in Western Europe.[5] In Russia itself, graft hybridization had been incorporated into the plant breeding program of Ivan Vladimirovich Michurin. Nicknamed "the Russian Burbank" in reference to the American horticulturalist Luther Burbank, Michurin had been a pioneering fruit breeder, whose authority was effectively co-opted after his death by Lysenko and the Soviet state (for an example of one of Michurin's grafts, see fig. 9).[6] Outside of the Soviet Union, graft hybrids were viewed as one of the most compelling aspects of Lysenkoism. Scientific organizations, such as the Genetics Society of America, were loath to involve themselves in the Lysenko affair.[7] Individual biologists therefore took up the fight, with the ambiguity surrounding the existence of graft hybrids becoming a recurring theme. When challenged to "provide minimal empirical support of Lysenkoism," biologist and science writer J. B. S. Haldane pointed to graft hybridization.[8] Most Western scientists, however, were unconvinced. Reacting to reports of the 1948 Academy of Agricultural Sciences meeting, American geneticist Robert C. Cook argued that Lysenko's graft hybrid plants were no different "from earlier chimeras of this kind reported in the literature of 'reactionary' biology."[9]

To explore whether Lysenko's embrace of the graft hybrid afforded him greater legitimacy in the West, I focus attention on Britain, home to a group of left-leaning scientists, shaped by their experiences of the First World War and appalled by the rise of fascism.[10] Two communist biologists are of particular interest. J. B. S. Haldane and his junior colleague Angus John Bateman were practicing geneticists and members of the Communist Party of Great Britain. Both had also worked at the John Innes Horticultural Institution, long home to experiments on grafting and graft hybridization. Haldane's participation in the Lysenko affair is often remembered via a disastrous BBC radio debate, where he equivocated on the legitimacy of Lysenko and concealed the death of the agronomist and geneticist Nikolai Vavilov at the hands of Soviet authorities.[11] Haldane was a complex character. His admiration for certain aspects of Soviet biology was tied up with his concerns for global food equity and security.

FIGURE 9. Sunflower grafted onto pear. Interspecies grafting was an important part of the plant breeding program of Russian horticulturalist Ivan Vladimirovich Michurin. Idolized by the Soviet state, Michurin would be used by Lysenkoists to support their beliefs. He would also be cited by Western biologists sympathetic to Soviet science as a more palatable alternative to Lysenko. I. V. Michurin, *Itogi Shestidesiatiletnikh Rabot* (Moscow: Ogiz RSFSR, Poligrafkniga Trust, 1934), 35.

Rather than using graft hybridization as a defense for Lysenko, he preferred to withhold judgment until more scientific data arrived from the Soviet Union. Bateman, on the other hand, was dismayed by the party's embrace of Lysenko. He decided that Lysenko's pronouncements on graft hybrids could not be trusted. In the mid-1950s he attempted to replicate graft hybridization experiments from communist Hungary and declared them lacking.[12]

Looking beyond the realm of the professional scientist, mid-twentieth-century Britain was home to an array of institutions and organizations dedicated to supplying farmers and horticulturalists with the latest information on breeding techniques. Two members of the Imperial Bureau of Plant Breeding and Genetics reviewed Lysenko's biology and withheld judgment on whether graft hybridization was possible. Sympathetic or neutral attitudes to aspects of Soviet biology existed in other organizations, even after Lysenko's fall from grace in the Soviet Union. During the 1950s and early 1960s, the Commonwealth Agricultural Bureaux promoted graft hybridization across the crumbling British empire. In 1967 British fruit grower Ben

Tompsett embarked on a cultural exchange visit to the Soviet Union. Here he was shown flourishing orchards in inhospitable regions: an environmental transformation supposedly made possible through the power of grafting. The farmers and horticulturalists who worked their fields and orchards firsthand were far more open to Lysenko's ideas than their scientific counterparts were. Despite the fallout from the Lysenko affair, the graft hybrid remained a plausible tool to upset the Western understanding of inheritance. As poorly accepted as Lysenko was in the West, the graft hybrid was the most well received—or at least irrefutable—part of his biology.

Haldane and the Graft Hybrid

One of the most vocal and persistent defenders of Lysenko and the wider communist project in Britain was John Burdon Sanderson—better known as J. B. S.—Haldane, a renowned biologist and popularizer of science. Haldane was introduced to the science of genetics earlier than most. In 1901 his father, the physiologist John Scott Haldane, took him to the Oxford University Junior Scientific Club. Here, the eight-year-old J. B. S. Haldane listened to a lecture on Mendel's laws by the biologist Arthur Dukinfield Darbishire.[13] Despite an early start, many years and trials would follow before the young Haldane would return to genetics. He assisted his father—occasionally as a human test subject—in mining and diving experiments. Haldane recalled that by age thirteen he was carrying out calculations for his father, and he presented his first paper to the Physiological Society at seventeen. Outside of his close family, a good scientific education was harder to come by. Upon leaving school, Haldane attended New College, Oxford, where he took a course on literature and humanities nicknamed the "Greats." Its dispassionate analysis of ancient history and philosophy, however, came at a moral cost. "The successful Greats Man," Haldane would later recall, "with his high capacity for abstraction, makes an excellent civil servant, prepared to report as unemotionally on the massacre of millions of African natives as on the constitution of the Channel Islands."[14]

With the outbreak of the First World War, Haldane found himself fighting on the front line. However, he did not allow this minor inconvenience to derail his scientific interests. He had recently encountered Darbishire once again, this time through the latter's papers on mouse breeding and Mendelian genetics. Darbishire had kept detailed notes and exquisite illustrations, which traced the appearance of different coats and eye colors through the generations of his mice. He had noted, however, that the characteristic

Mendelian ratios did not seem to appear. Haldane worked with his friend A. D. Sprunt (who was later killed in action) and his sister, Naomi Haldane, to interpret Darbishire's extensive data. They effectively saved the applicability of Mendelian genetics by uncovering evidence of "genetic linkage" (when genes next to each other on the chromosome are inherited together).[15] Haldane's soldiering was also admired but was occasionally less than successful. While acting as a platoon commander in the Black Watch Regiment in the spring of 1915, he was ordered to fall back and assist his father in developing a respirator to counter the German use of chlorine gas. Several painful experiments later they produced a working respirator. At this point Haldane received news that his regiment was poised to attack the German lines. Wishing to join them, he attempted to run back to the front, only to find that the chlorine had restricted his airways. "The best I could achieve," he lamented, "was a moderate trot worthy of an old gentlemen with chronic bronchitis."[16] Haldane missed the assault, was hit by enemy artillery fire, and ended up in the hospital.

When the war ended, Haldane threw himself back into genetic research. Drawing heavily upon mathematics and statistics, he gleaned further insights into genetic linkage and participated in the twentieth-century synthesis of Darwinism and Mendelian genetics.[17] During the late 1920s Haldane would regularly leave New College to visit the John Innes Horticultural Institution, where he had been appointed assistant director. Here he would meet a promising young geneticist named C. D. Darlington. He would later fall out with Darlington, who would assume the directorship of the John Innes when Haldane became professor of genetics at University College London in 1933.[18] Haldane's impact at the John Innes was limited, although he did conduct further work on genetic linkage in partnership with Dorothea de Winton, a professional gardener who had worked her way up in the John Innes to become officially recognized as a geneticist.[19] During this period, Haldane's politics began a gradual shift toward communism. He was unimpressed by the attitude of the "British establishment" toward science. Compared to their counterparts in the Soviet Union, complained Haldane, British politicians, generals, and bureaucrats neither understood nor appreciated science.[20] In 1928 he visited the Soviet Union and met Vavilov, who would later become a victim of the Lysenko affair.[21] By 1937 Haldane's first wife, Charlotte Franken, had become a member of the Communist Party of Great Britain. Although Haldane was not yet a member, he was a Marxist and a public supporter of the party.[22] He would later "dedicate" his unpublished memoirs to "Adolf

Hitler and Benito Mussolini, who converted me to cooperation—after softer arguments had failed."[23]

By the time the Lysenko affair erupted, Haldane had invested both his time and emotion in the communist cause. In 1942, Charlotte had returned from a visit to the Soviet Union. She told J. B. S. that Soviet society was now repressed and corrupt. Vavilov was missing, with rumors that he had been arrested and executed by the state. Haldane refused to believe her, and their divorce followed shortly thereafter.[24] Lysenko's biology represented a tricky diplomatic problem for Haldane. On December 27, 1944, Haldane received a letter from Lysenko through the Russian embassy in London. Lysenko expressed his "highest interest" in Haldane's *New Paths in Genetics* (1941) and provided a copy of his own book on heredity, "which in many points differs essentially from the orthodoxal mendel-morganistic genetics." When Haldane replied in January 1945, he probed Lysenko for more information. "Unfortunately," he wrote, "it is extremely difficult in this country to obtain accounts of the actual experimental results on which your views are based, and I should be very grateful if I could obtain some reprints giving accounts of the facts in question."[25] These reprints, if they existed, were not forthcoming. Haldane did not receive any further correspondence from Lysenko. He would spend the remainder of the Lysenko affair attempting to defend a strange biology on which he had no more information than many of his peers.

When Haldane attempted to defend Lysenko's biology, graft hybrids took center stage. In his 1947 book *Science Advances*, a series of essays originally published in such outlets as the communist *Daily Worker*, Haldane argued that Lysenko's attacks on genetics were nothing special. "Professor Jeffrey of Harvard," noted a finger-pointing Haldane, "has attacked genetical science much less temperately and on much flimsier evidence than Lysenko." He also pointed to the neo-Lamarckian activities of Ernest MacBride, whom we previously met discussing Paul Kammerer's career. Haldane depicted Lysenko as much more sophisticated than he probably was, while suggesting that the Soviet assault on genetics was directed against its overly simplistic teaching. He did not have many examples of where he thought Lysenko had gotten biology right. Yet graft hybridization did make an appearance: "He [Lysenko] was quite right in saying that so-called pure lines of plants are generally mixtures, and that an exact three to one ratio in accordance with Mendel's law is very rarely obtained. He also stated that in tomatoes and related plants a number of characters described as hereditary can be propagated by grafting. In just the same way [Clarence Cook] Little, [John Joseph] Bittner, and other workers at Bar Harbor, Maine, found that the tendency to breast cancer in mice, formerly regarded as hereditary, was largely transmitted through the mother's milk."[26] Here, Haldane was carefully making the case that grafting could pass certain characteristics from one plant to another. He did not, however, suggest that these characteristics were hereditary to begin with. This stance admitted that grafting was potentially a useful economic tool without admitting that true graft hybrids existed. Shortly after this passage, Haldane repeated Lysenko's argument that plant and animal breeding was often carried out without reference to genetics and that "geneticists have made exaggerated claims for the economic value of their science." Haldane concurred with Lysenko but argued that the "economic value of genetics is greater than he thinks." Concluding his thoughts on Lysenko's biology, Haldane stated that Lysenko had gone too far, both in his attacks on Thomas Hunt Morgan's chromosome theory and "in his claims concerning the possibility of transferring characters by grafting."[27]

Despite his misgivings about Lysenko, two aspects of Soviet science in general appealed to Haldane. One was its focus on practical questions, which he thought drove science toward matters of economic importance while allowing enough leeway for basic research. Another was what Haldane termed "the historical angle." Science conducted under the philosophy of dialectical materialism, such as Vavilov's research into the origin of crop plants, could shed new light on human history.[28] Both food security and the origin of agriculture had been on Haldane's radar for a long time. In a 1936 talk he had compared the domestication of plants and animals to the invention of the steam engine and the Industrial Revolution as milestones in human history. Never one to dodge controversy, however, Haldane saved a few scathing remarks for those who still believed that early agriculturalists had lived in a "natural" state. "A farm," he declared, "is as unnatural as a motor bus or a blast furnace." Ancient farmers cleared trees and drained land, spending their time looking after "most unnatural plants such as wheat and potatoes," or "unnatural animals like cows." He wished more people would recognize that civilizations were made up not just of their human inhabitants but also of domesticated plants and animals. Remembering just how profound the invention of agriculture had been could help his contemporaries grasp how machinery was changing modern society, hopefully leading to closer cooperation between industry and agriculture.[29]

By the mid-twentieth century, the modernization of agriculture became a profound moral and scientific project for Haldane. This included the production of synthetic food. He predicted that all food would one day be synthetic, as "in the remote future men and women will have so much respect for life that they would no more think of eating a dead animal than a dead man, and will prefer not even to kill plants for food." Grass proteins, cultured algae, and fat produced from coal could all help achieve this vision. Haldane's vision for agriculture soon began to tie into his politics. The British government, he declared, would not embrace his drive for synthetic foods as it "has spent so much money on ill-advised schemes of Colonial development."[30] This obstruction of science was a travesty. "Give us biologists a free hand with food production," declared Haldane, "and we will give you a world in which there is enough food for everyone."[31] In another essay, he argued that people were becoming increasingly disillusioned with science thanks to global hunger. Tackling starvation was ignored by British science policy in favor of less palatable alternatives. "We spend colossal sums on research designed to give us new weapons," bemoaned Haldane, "and very large ones on research on the production of new luxuries such as television and nylon."[32] Furthermore, continued Haldane, a transformation in food production could not be led by private enterprise, as "it would not pay."[33]

Against this backdrop of frustration with the British government and its inaction on global food security, radical alternatives such as graft hybridization began to look increasingly attractive to Haldane. An opportunity to reevaluate the graft hybrid came after the 1948 International Congress of Genetics in Stockholm, Sweden. Lysenko came under fire at the congress. Its participants noted the suspicious absence of Vavilov and listened to an address by the geneticist Hermann Joseph Muller. Muller had been driven from the Soviet Union during the Lysenko affair, subsequently winning scientific plaudits for his work inducing mutations with radiation.[34] Haldane was smitten with this idea, and described how German geneticist Charlotte Auerbach, while a refugee in Edinburgh, had induced mutations using mustard gas and other chemicals. "This means," concluded Haldane, "that we have a series of agents with which we can make species more variable, and later select those variants which we desire." Such claims, it seemed to him, were somewhat like those espoused by Lysenko. "None of the speakers made such wide claims as Lysenko has done," he noted, "and I think it will be found that some of Lysenko's claims have been exaggerated." On the other hand, mused Haldane, "a number of workers are now nearer to Lysenko's point of view than they were ten years ago."[35] He also noted that genetics continued to struggle with hybridization, prescribing "dialectical thinking and practice" to overcome these difficulties. "In the Soviet Union," wrote Haldane, "grafting as well as hybridization have been used. Unfortunately full details are hard to obtain in this country."[36]

What are we to make of Haldane's lukewarm support for graft hybridization during the 1940s? The mathematical side of Haldane was clearly perturbed by the absence of hard, numerical data from the Soviet Union. Nevertheless, the graft hybrid possessed a certain allure. It was a radical form of plant breeding for a world wracked by hunger and dominated by food conglomerates. Marxism, which Haldane embraced in part for its focus on the practical, also seemed to be a natural fit with the relatively simple and experimental nature of graft hybridization.[37] Away from his political opponents, however, Haldane was much more critical of Lysenko's biology. A crunch point for him came in November 1948, when the Engels Society attempted to issue a statement on Lysenko and Soviet biology at the request of the Communist Party of Great Britain. The society was little more than an unofficial "discussion club" on the Marxist philosophy of science, but did include some intellectual heavyweights.[38] Its statement on Lysenko was drafted by the botanist A. G. Morton. It reaffirmed the inseparable connection of biology with agriculture and railed against August Weismann's work. One of the ways in which the heredity of plants could be altered, stated Morton, was by "grafting different varieties together and creating graft hybrids."[39]

Haldane was furious. In a letter to Maurice Cornforth, the Marxist philosopher who led the society, he demanded that a planned meeting be pushed from November to December. "I protest in the very strongest manner against any statement being produced on the subject of Lysenko," wrote Haldane, "until the matter has been discussed by the geneticists who are members of the Society, and I trust no such meeting will be held." He ended by threatening to resign his membership.[40] Lysenko was a sensitive subject for Haldane at this point. He had always tried to walk a fine line during the Lysenko affair, promoting what he saw as the positives of Soviet biology while downplaying some of Lysenko's more outlandish claims. This line had come under intense scrutiny following Haldane's disastrous BBC debate a few months earlier. Likely aware of this context, Cornforth was reconciliatory in his response. He postponed the Engels Society meeting and reassured Haldane that he had no intention of bypassing the opinions of its geneticists.

A meeting of the Engels Society geneticists took place on December 15, 1948. Judging by Haldane's preparations, it was probably a lively one. A discussion statement, "In Support of Lysenko," was issued to participants beforehand.

Haldane littered his copy with question marks and the occasional blunt "no." Most of his complaints were directed at unflattering characterizations of contemporary biology. He left a question mark next to the statement that scientists separated from the working class "had neither opportunity nor incentive to wrest ever more from nature for a society that dreaded conditions of over-production as disastrous."[41] Other claims brought a more forceful response. Haldane scrawled "no" over Lysenko's definition of heredity as "the property of a living body to require definite conditions for its life, and to respond in a definite way to various conditions," which the society was attempting to label "accurate and complete."[42] Haldane was less combative regarding the practical results of Soviet biology. The Engels Society statement explained how the "physiological treatment of the organism (graft hybridisation, partial vernalisation, distant hybridisation etc.)" had led to results that could not be accounted for by Mendelian genetics. Haldane wrote "Evidence" next to this claim but did not contest the following statement, that "the successes of Michurin in the field of horticulture have been repeated and extended in all spheres of agriculture by Lysenko and other Soviet biologists."[43]

Haldane allowed his membership of the Communist Party of Great Britain to expire in 1950. But distancing himself from the political controversies of the Cold War did not imply that he had given up on grafting. In a 1955 essay, colorfully titled "Some Alternatives to Sex," Haldane finally went into some detail on graft hybrids. He acknowledged that grafting had an important horticultural role, uniting the properties of stock and scion from different plant species or varieties. He also had few reservations in supporting the claim of Soviet biologists that "the material basis of many characters can be transmitted, and that the germ cells of an individual altered by grafting can transmit them." Plant viruses, noted Haldane, could explain this phenomenon, as they moved across grafted plants and altered their genes.[44] On the more controversial question of whether genetic material could be directly transferred between plants—without an intermediary like a virus—Haldane pointed to the work of Hungarian botanist L. J. M. Felföldy. Felföldy had apparently grafted two varieties of tomato, which differed from each other in key traits determined by their genes. When Felföldy bred the grafted plants with themselves (self-fertilization), their offspring possessed a mix of traits from both varieties. Haldane, however, was still not convinced that it was necessary to invoke graft hybridization:

> Allowing for some uncritical work, I find it extremely hard to suppose that all results of this type are due to fraud or technical error. If not, the most

likely hypothesis seems to be that we are dealing with transduction by a "virus." The Solanaceae, the order to which tomatoes belong, harbour many "viruses," of which some are harmless to their normal hosts. It will be of extreme interest to discover whether the agents which have passed from one plant to another proceed to settle down and behave as genes with normal linkage relations, as the temperate phages do in some bacteria.[45]

By invoking viruses as carriers of genetic material, just like the "molecular scissors" used in gene editing today, Haldane could explain the appearance of graft hybrids in the Soviet Union in a way that was palatable to Western geneticists. He concluded as much in his 1955 essay. Nonsexual inheritance, wrote Haldane, "no more disproves Mendelism or Mendel-Morganism than the existence of parthenogenesis disproves sex, or the existence of magnetism disproves gravitation."[46] Others, however, would disagree. Haldane never actually carried out his own experiments on graft hybridization. His friend and colleague Bateman did. An active member of the Communist Party of Great Britain, Bateman would become an outspoken anti-Lysenkoist within its ranks.[47] Graft hybrids did not only have the power to sweeten the bitterness of Lysenko's biology. As the case of Bateman would show, their refutation also had the power to considerably undermine Lysenko.

The Skeptical Communist

Trained as a botanist at King's College, London, Angus John Bateman spent the mid-twentieth century moving across the landscape of British genetics and agronomy. He worked on everything from cross-pollination and quantitative genetics to grafting and animal breeding.[48] His Marxist politics brought him into contact with Haldane when Bateman was in his early twenties. An initial letter from Bateman to Haldane, written possibly as early as 1938, made their political alignment more than explicit. Addressing Haldane as "Comrade," Bateman asked Haldane for career advice. "I feel I can get a more satisfactory answer from a Marxist," wrote the young botanist, "than I have so far got from my College Dons."[49] Intriguingly, Bateman was thinking of acquiring a letter of recommendation from Reginald Ruggles Gates, whom we earlier encountered flirting with the graft hybrid in the run-up to the First World War. "If you know Prof Reginald Ruggles Gates' politics," wrote Bateman, "you may be able to tell me whether it is advisable to appear as a 'Red' before him and his staff." Bateman concluded by asking, "Can a non-mathematician be a geneticist?"[50]

Haldane must have been encouraging, or at least given Bateman an accurate reading of Gates's politics, as their correspondence continued. Haldane suggested that Bateman investigate the genetics of *Primula elatior*, or oxlip, a flowering plant found in woods and meadows.[51] To Bateman's great embarrassment, this venture was an "ignominious failure." He had read that the plant did not flower until May and did not encounter any specimens in Oxfordshire or Surrey over the Easter break. Bateman therefore postponed a planned hunt for the flower, only to find that he was beaten to the punch. On returning to King's College on April 25, he found that "a demonstrator who had just come from Cambridgeshire" had brought specimens of *Primula elatior* and its hybrid forms to display at the college.[52] Bateman's embarrassment was temporary, with the young botanist finding himself at the John Innes Horticultural Institution in 1942. Here, he worked with *Drosophila* fruit flies and in 1948 conceived of "Bateman's principle," that variations in an individual's reproductive success are supposedly greater for males than for females. Lysenko's address to the Lenin Academy of Agricultural Sciences occurred the same year. As a Marxist geneticist in a British horticultural institution, how would Bateman respond?

Bateman's public response to the Lysenko affair would have been enough to make even Haldane blush. In a 1949 review of Lysenko's address, published in the journal *Scientific Worker*, Bateman accused the "press and radio in Britain" of creating the false impression that the Soviet Union had "decided to take their money off Mendelism and put it on Michurinism," based not on science but on purely "political motives." He portrayed the Lysenko affair instead as a clash between academic genetics and practical agronomy. Mendelians dominated Soviet universities and research institutions, claimed Bateman, while the Michurinists were more numerous in agricultural research stations. The latter were admittedly guilty of ignoring the accumulated experimental evidence behind genetics. Yet Mendelians, proclaimed Bateman, had not adequately applied themselves to breeding or addressed the claims of Michurinists. Among these claims was graft hybridization. "The technique of grafting as a method of plant breeding," declared Bateman, "greatly developed by Michurinists, is being increasingly used in non-Soviet countries."[53] Bateman's defense of Lysenko provoked a savage backlash from Darlington, director of the John Innes Horticultural Institution and Haldane's rival.[54] In a letter to Haldane, Bateman reported that Darlington was "livid," and "if he had had the power, he would have sacked me on the spot." According to Bateman, Darlington demanded that

the Agricultural Research Council, the main government body for the funding of agricultural science, remove Bateman from the John Innes on grounds
of incompetence. Luckily for Bateman, however, Darlington had previously
criticized the council. It is possible that they therefore took this opportunity
to ignore his complaints.[55]

Away from the public spotlight, however, Bateman had serious concerns
about Lysenko. On August 29, 1948, he expressed these to Haldane in a
private letter. After the 1948 International Congress of Genetics, Bateman
was left with "a very guilty feeling that I had failed in my duty as a Communist." He was disappointed that communist geneticists at the congress
had not taken the opportunity to discuss Lysenko. "The Lysenko heresy
has been going on for a decade," he reminded Haldane. "Here was a golden
opportunity for a decisive formulation of policy on the relation of genetics
to dialectical materialism and exposure of the anti-scientific mumbo-jumbo
of Lysenko and his school."[56] Bateman was angry that scientific attacks on
Lysenko, whom he described as being raised to a "god-head," were construed
as political disloyalty. This situation only helped people like Darlington,
who wished to portray Soviet science as based solely on political ideology.[57]
Bateman concluded that communist geneticists had no choice but to collec-
tively evaluate the merits of Lysenko's case against genetics. He requested
that Haldane arrange this meeting.[58] The attempt to issue a statement on
Lysenko and Soviet biology a month before the geneticists had had the
chance to meet in December 1948 was the reason Haldane almost fell out
with the Engels Society.

Bateman had much to debate at the meeting of communist geneticists
on December 15, 1948. He admitted in his discussion statement that there
was too much about Lysenko to cover in a single sitting. Bateman instead focused on Lysenko's attacks on genetics and the evidence behind them. These
attacks were not only on the scientific basis of genetics but included charges
that the subject had become "divorced from practice" and had transformed
"plant breeding into waiting for something to turn up."[59] Bateman, still based
at the John Innes Horticultural Institution, thought that the relationship
between genetics and agriculture—or lack thereof—was the primary reason that the Soviet Union had rejected the science. Lysenko's refutation of
genetics was based on a few key bodies of evidence. These included, wrote
Bateman, "the effects of stock on scion and vice-versa in graft hybrids *and
the inheritance* of those effects." Bateman had already made up his mind about
what he wanted the meeting to achieve. Marxist geneticists, he hoped, would

expose the "unsoundness of Lysenkoism, the dangers of placing too much reliance upon him, [and] the undesirability of suppressing genetical teaching and research in the U.S.S.R."[60]

A summary of the meeting was later written up by Bateman and circulated to the Engels Society in January 1949. It was a combination of carefully targeted criticism toward Lysenko and a promotion of the flexibility of genetic theory. To begin with, wrote Bateman, Lysenko's attacks on Mendel and Morgan were misplaced. These figures represented an early incarnation of the science, which modern geneticists had moved on from. In fact, declared Bateman, "Modern genetical theory is inevitably more in accord with dialectical materialism than the necessarily crude early theories."[61] Genetics, asserted Bateman, did not claim that heredity was completely independent of the environment. Yet there were several points where Lysenko and modern genetics were at odds. These were the inheritance of acquired characteristics, the Lysenkoist claim that competition did not occur between individuals of the same species, and graft hybridization. Bateman and his fellow geneticists flatly denied the assertion that there was no competition within a species. On acquired characters and graft hybrids, there was less certainty.

On the inheritance of acquired characters, reported Bateman, many of the experiments in the Soviet Union were conducted "without a very detailed examination of the results, without the use of proper controls, [and] without valid statistical tests." Graft hybridization in the Soviet Union suffered from the same shortcomings: "A similar attitude is adopted by geneticists to the experiments with the inheritance of characters transmitted by grafting. Isolated cases of this are well substantiated and geneticists have a perfectly valid explanation, that they are due to cytoplasmic genes which are transmitted through grafts and thence to the sex cells."[62] This passage was likely a reference to the transmission of genes outside the nucleus, such as those in plant viruses, which would later be seized upon by Haldane as an explanation for graft hybridization. Bateman also reported that the communist geneticists were generally less impressed by the claim that grafting and other Lysenkoist methods had revolutionized agriculture in the Soviet Union. "Michurin and others," he wrote, "claim wholesale effects [from grafting] which could be used as an organised system of plant breeding of comparable if not of greater importance than sexual hybridisation. This is in conflict with all other experience with grafting." For now, the communist geneticists would once again "reserve judgment" pending the release of more detailed results from the Soviet Union.[63] Although the meeting did not result in a public denouncement of Lysenko, there was apparently little disagreement with Bateman's critique. There were also some minor shifts within the British Marxist community. In a letter to Haldane on February 9, 1949, Bateman expressed his delight that his discussion statement had changed the mind of the Marxist economist Emile Burns. "He was quite satisfied with my draft," effused Bateman, "and was even prepared to admit errors in Cornforth's criticism. The reason? He has read an elementary book on Cytology! He says he finds the controversy much easier to follow now."[64]

When news of Felföldy's tomato-grafting experiments reached Britain, Bateman became directly entangled in the graft hybrid controversy. Based at the Biological Research Institute in Tihany, Hungary, Felföldy had grafted a tomato variety called Oxheart onto another tomato named Golden Apple. The grafted Oxheart type differed from Golden Apple in five ways, each based on a known Mendelian gene. A Mendelian dominant gave Oxheart tomatoes red flesh, while a recessive gave its fruit colorless, instead of yellow, skin. Other genes dictated whether the plant gave long or tipped fruits, or if these tomatoes were divided into many chambers instead of the traditional two or three. Felföldy did not notice any changes on the grafted plants themselves. But when he self-fertilized the grafted scion of Oxheart, some of its seedlings showed a real mix of characteristics. These included tomatoes with red flesh and yellow skin, with shortened fruit and an intermediate number of chambers. The charges of inadequate data could not be leveled against Felföldy. He had also obtained seedlings with mixed characteristics from his Golden Apple stock, while finding only plants with the expected characters in stocks of nongrafted Oxheart and Golden Apple kept as controls. Felföldy grew another five generations of plants from his hybridlike seed, which began to segregate into Mendelian ratios.[65] It seemed that Mendelian genes had moved between the two grafted tomato varieties and were inherited by subsequent generations.

Bateman was tasked with repeating Felföldy's experiments by Dan Lewis, a pomologist who had become head of the Department of Genetics at the John Innes in 1949. In 1954 Lewis and his colleague Lesley Crowe had bred a mushroom called *Coprinopsis cinerea* to "demonstrate cytoplasmic inheritance."[66] Lewis was therefore open to Bateman working on alternative forms of heredity. If what Felföldy had said was true, then germplasm could be shared and inherited from grafted plants. Bateman seized this opportunity, noting that "a closer parallel than hitherto should have been possible between Michurinist and Western experimenters" had emerged.[67] He wrote a letter to Felföldy about his intention to repeat Felföldy's work. The Hungarian botanist welcomed the scrutiny, providing Bateman with seeds from his own stocks of Oxheart and Golden Apple tomatoes. Bateman set to work. He cut slots in the stems of his stock plants, then trimmed the stem of his scion into a wedge and inserted it in the slot. Besides this standard grafting procedure, Bateman also took some more unusual actions. A week after his plants had been grafted, he defoliated some of the scions and added a new set of grafts with their leaves removed. Bateman explained that these modifications "should increase the physiological dependence of scion on stock and thus, according to Michurinist principles, increase the yield of graft hybrids."[68] Despite this intervention, Bateman found no evidence of hybrid characteristics in his grafted plants.

If graft hybridization was not possible between these tomato varieties, then what had Felföldy observed in his greenhouse at Tihany? Bateman noted that one of the tomato plants he grew from Felföldy's seeds turned out to be a first-generation hybrid of Oxheart and Golden Apple. "If such a plant had been inadvertently used as scion in a graft-hybrid experiment," wrote Bateman, "some of Felföldy's results could be explained. Though I found no cases of spontaneous cross-fertilization in my material, Hungarian conditions are evidently more conducive to it." He concluded that many of Felföldy's graft hybrids were "environmentally produced variants" or had emerged from accidental cross-pollination. Bateman admitted that he could not account for the intriguing characteristics of some of Felföldy's tomatoes. He therefore argued against graft hybridization on grounds of probability. Felföldy, noted Bateman, had managed to obtain fifteen plants out of eight, which, "if we do not accept as genuine vegetable hybrids, are difficult to account for." His experiments had found no hybrids from more than one thousand plants. On these grounds, Lysenko's claim that graft hybridization could replace traditional hybridization as a plant breeding technique seemed unrealistic. Bateman concluded that as his own experiments had "provided more favourable conditions for vegetative hybridization" and still come up with nothing, "one cannot take it [graft hybridization] seriously."[69]

One of William Bateson's mantras was to "treasure your exceptions," a lesson that Bateman seemed to forget. He instead dismissed graft hybridization as a botanical phenomenon on the grounds that it was unlikely to occur. Part of his argument for doing so was that his experiments had replicated some of the techniques suggested by Michurin for increasing the likelihood of graft hybridization. While some of his grafted plants "were in all material respects the same as Felföldy's," the majority seemed to be different.[70] Bateman had essentially tried to kill two birds with one stone by simultaneously testing Michurin's and Felföldy's results. If Felföldy had found a genuine means of producing tomato graft hybrids, Bateman would have drastically reduced his own chances of doing so by altering many of his tomato plants. Bateman wrote up his findings in February 1955, only to receive more results from Felföldy shortly thereafter. Felföldy had raised another generation of plants from the seeds of his graft hybrids, which led to more unexplained results. Bateman admitted that some of the Hungarian tomatoes were not actually environmentally produced variants but may have been polymorphic or somehow altered by grafting. Bateman did not shift from his original stance that graft hybrids should not be taken seriously.

Two years later, a published response from Felföldy confessed that an experimental error could account for some of his graft hybrids. Felföldy had been alerted to the presence of a sexual hybrid in his plants by both Bateman's article and personal correspondence. "I can only explain the occurrence of these sex-hybrids as due to lack of care," wrote Felföldy. "In that summer there were in the greenhouse many tiny insects, including thrips, and the simple isolation with cotton"—that is, covering the plants with cotton bags to prevent cross-pollination—"must have been insufficient." Elsewhere, Felföldy pushed back against Bateman's conclusions. He did not accept that his tomatoes could have been polymorphic, since the varieties had been bred in Tihany since 1946 with "neither spontaneous cross-fertilization nor mutation-like changes." Grafting had changed the appearance of these plants, but these changes had not been inherited by subsequent generations. "These non-segregating types," admitted Felföldy, "have no practical importance."[71] Unfortunately for the graft hybridizers, the most striking results from Hungary may well have been down to the unwanted activities of a few troublesome insects. Although kept separately and confined within their cotton bags, at least a few of Felföldy's plants had exchanged their genetic material prior to grafting. As the saying goes, life finds a way.

One of the most intriguing aspects of graft hybridization in the mid-twentieth century was how a simple horticultural technique became a political football. As the case of Bateman shows, it was not simply that communist biologists lined up in support of the graft hybrid. It was Bateman's drive to oust Lysenko from the intellectual life of British communism that brought him into repeated contact with graft hybridization. Bateman's political agenda reappeared when he was asked to recreate Felföldy's grafted tomatoes.

Rather than simply doing so, he took the opportunity to test several aspects of Michurin's biology. He quickly found them lacking and dismissed the possibility of graft hybridization. Haldane, however, sought a compromise. Felföldy's 1957 response to Bateman noted that Haldane had suggested that "the inheritance of characters [through grafting] may be attributable to virus transduction."[72] Here, then, we have two communist biologists at odds. Both were geneticists and both had worked at the John Innes Horticultural Institution. Given their proximity to horticultural practice, we will now ask what British breeders and farmers made of the Lysenko controversy. Were they open to the existence of graft hybrids?

A British Fruit Grower in the USSR

"Dog breeding is full of pitfalls and dog breeders are full of folklore," raged C. D. Darlington, director of the John Innes, in 1953.[73] Darlington's target was a recent monograph on dog breeding by Marca Burns, a member of the Commonwealth Bureau of Animal Breeding and Genetics at the University of Edinburgh's Institute of Animal Genetics. Burns had not shied away from addressing the Lysenko controversy, remarking that the "most controversial question in all the history of the science of genetics has been whether or not peculiarities acquired by an individual during its lifetime can be passed on to its descendants."[74] She claimed that British breeders of livestock and working dogs had never been disabused of the notion that acquired characters could be inherited. The theories of Lysenko had therefore "aroused much interest among farmers."[75] Darlington was outraged that such views were "still held by illiterate people all over the world" and were being expounded, "coupled with the name of Lysenko," by a division of the state-funded Commonwealth Agricultural Bureaux.[76] He claimed to be unsurprised that this had occurred in unorthodox Edinburgh. As we have seen, Francis Albert Eley Crew—director of the Animal Breeding Research Department and founder of the Institute of Animal Genetics—had written in support of non-Mendelian means of heredity back in the 1920s, using graft hybrid guinea pigs as experimental evidence.[77] In Britain, as in much of the Western world, Lysenko's theories were generally met with skepticism. Yet was Darlington, himself a figure of great controversy, on to something? Were members of state-funded agricultural organizations, along with the farmers and breeders they claimed to represent, secretly sympathetic to Lysenko? An early examination of Lysenko's biology was produced by two members of the Cambridge-based Imperial Bureau of Plant Breeding and Genetics, P. S. Hudson and R. H. Richens, in 1946. Although they had professed to have produced an impartial assessment of Lysenkoism, Hudson and Richens were generally damning. The scientific claims of Lysenko, including on the vernalization of wheat, were dismissed as either false or completely overblown. One exception, however, was graft hybridization. "The controversy as to whether or not graft hybridization occurs," they wrote, "is one of the several long-standing problems of biology." They knew that Frederick Ernest Weiss and William Neilson Jones had investigated the phenomenon during the 1930s only to take the neutral stance that graft hybridization was possible but had not yet been demonstrated.[78] After 1940, Hudson and Richens claimed, "evidence on the anomalous effects of grafting has come to constitute the principal experimental basis for Lysenko's theories." They speculated that this shift had come about thanks to the absence of compelling empirical evidence to support other aspects of Lysenko's new biology.[79]

Of particular interest to these two members of the Imperial Bureau of Plant Breeding and Genetics was a series of tomato grafting experiments, published in Russian in 1941 by A. A. Avakjan and N. G. Jastreb. These experiments, wrote Hudson and Richens, had come to represent "one of the corner-stones of Lysenko's system."[80] Avakjan and Jastreb had grafted two varieties of tomato, one with large yellow fruits and one with small red fruits. They found that the grafted plants bore fruit in a dizzying array of colors. Some of their tomatoes were red, some yellow, others were pink or yellow with pink stripes. The seeds of these tomatoes also gave them plants with a mixture of colors. Far more importantly, the second generation of tomatoes—the offspring of the offspring of the original grafted plants— maintained a hybrid appearance, with red, yellow, pink, and striped fruit. If true, these results suggested that the colors of the original tomato plants had successfully crossed the graft junction and had been inherited through multiple generations. The tomatoes must have been true graft hybrids, not chimeras. "It cannot be denied," wrote Hudson and Richens, "that their results are somewhat unexpected as they stand." Results of such importance, they continued, required the utmost scientific rigor. Hudson and Richens complained about the presentation of the Lysenkoists' data and their lack of controls. They also noted that it was not until 1943 that Lysenko had advised his disciples to cover flowers with gauze bags to avoid accidental cross-pollination. "Stray cross-pollination would certainly explain many of the results of Avakjan and Jastreb," Hudson and Richens concluded, "and until this and the other points militating against their results can be removed by confirmatory experiment, it is only possible to suspend judgment in the meanwhile."[81]

This neutral stance on graft hybrids mirrored that taken by Haldane. If Darlington had been aware of how many keen British breeders and members of agricultural institutions had approached Haldane during the Lysenko affair, he would have been even more apoplectic than usual. Burns, the author of the dog-breeding monograph, was just one member of Haldane's extensive list of correspondents. In December 1940 they had corresponded about a paper by Burns on greyhound breeding and in 1953 Haldane received a copy of Burns's book.[82] These contacts, however, did not necessarily equate to a widespread interest in either Lysenko or graft hybridization. John Monck, secretary of the British Cattle Breeders' Club, contacted Haldane on behalf of the club in 1948 to solicit his advice on genetics. "Many of us feel," wrote Monck, "that hitherto genetics [has] been too much of an academic and class room subject and not nearly closely enough allied to the practical problems of breeding our larger farm animals."[83] George M. Odlum, a member of the club, had told Monck that he sought Haldane's insight on "environmental differences" in dairy cow breeding and suggested that Monck seek "an American Professor to lecture us on the practical application of genetics to dairy cattle."[84] Mid-twentieth-century British farmers, like their early twentieth-century American counterparts, were more interested in results and cash flow than ideology or biological theory.

Thanks to Hudson and Richens, the graft hybrid had received a fair hearing at the Imperial Bureau of Plant Breeding and Genetics. Other breeding and horticultural organizations in Britain also addressed the Lysenko affair. In 1950 an English translation of Michurin's *Selected Works* was published by the Foreign Languages Publishing House in Moscow, with a review appearing in the journal of the Royal Horticultural Society the following year.[85] A self-taught and self-made fruit grower, Michurin was often cited by Lysenko and his followers as the basis of the new progressive Soviet biology. The Royal Horticultural Society's reviewer, M. B. Crane, noted that Michurin discussed the "influence of the stock upon the scion" and gave an account of his creation of a "so called vegetative [graft] apple-pear hybrid" in 1898. Crane was skeptical. "I cannot accept this as an example of vegetative hybridisation," he wrote, "nor can I find anything in the book which proves there is such a thing as vegetative hybridisation."[86] He dismissed Michurin's discussion of the inheritance of acquired characters as "mainly philosophical" and lacking proof. Yet to Crane's surprise, there were "some accounts in this book which most biologists, at least most of those outside Russia, can accept." Michurin did not deny Mendelian laws but sought to introduce various amendments and additions to them. For Crane, this inclusiveness aroused suspicion. "One is bound to ask," he wrote, "whether Lysenko or his colleagues have really been honest with Michurin."[87]

Leading members of the Royal Horticultural Society were unlikely to embrace Lysenkoism. An outsider's perspective on the relationship between the society and plant breeders was provided by M. J. Sirks, professor of genetics at the Government University, Groningen, in the Netherlands. The Royal Horticultural Society, claimed Sirks, had long been in the business of clearing up misunderstandings and falsehoods in horticulture. This business was continued, from a "genetical point of view" and to the "great credit" of the society, by botanist E. K. Janaki Ammal in the society's laboratories at Wisley Gardens.[88] Janaki Ammal was a regular contributor to the society's journal, where she promoted the chromosome and its benefits for plant breeders. "The number of chromosomes in a plant," she wrote in 1951, "their shape and other morphological characters seen during germ formation, tell us not merely what is going on in the living plant but also what has happened in the course of its history."[89] Janaki Ammal saw the manipulation of chromosomes using substances like colchicine as possessing vast potential for the improvement of "economic and ornamental plants": vital in a "world which is yearly finding it more difficult to feed itself."[90] Within the Royal Horticultural Society, genetics was an exciting and dynamic field of great potential. From this perspective, there was little or no need for such revolutionary and unproven alternatives as graft hybridization.

Animal Breeding and Genetics was not the only branch of the Commonwealth Agricultural Bureaux that might be accused of erring in favor of Lysenko. In 1963 Robert L. Knight, of the Commonwealth Bureau of Horticulture and Plantation Crops, based at the East Malling Research Station in Kent, England, issued a list of papers published on horticulture since 1900. Like Burns's dog monograph, this technical publication was intended to update breeders—in this case, breeders of common fruit trees—with information on "breeding, genetics and cytology."[91] So far, so orthodox. But Knight also included several papers that proclaimed the truth and utility of graft hybridization. These papers included several accounts of graft hybridization for plant-breeding purposes in the Soviet Union. A 1948 contribution by S. I. Isaev explained how the "root mentor effect" had been used to transfer "hardiness" across the graft junction to produce apple varieties suited for central Russia.[92] Experiments in Leningrad were reported as transmitting heritable changes from scion to stock in 1954.[93] The most recent contribution from the Soviet Union dated to 1960 and described the breeding of dwarfed apple trees using "vegetative hybridization."[94] Through such publications, public bodies did promote graft hybridization to British horticulturalists as a genuine means of breeding new plant varieties. Knight was lucky not to have his work spotted by an enraged Darlington.

The promises of Soviet agriculture provided another possible source of temptation for growers to embrace graft hybridization. Although Lysenko had fallen from grace after Nikita Khrushchev was deposed in 1964, grafting remained a major part of the Soviet horticultural scene for years. In May 1967 British fruit grower Ben Tompsett visited the Soviet Union under a cultural exchange agreement. Tompsett was immediately impressed. He reported that, despite having previously "visited the major fruit-growing areas of the world," he had nowhere before seen "development proceeding on such a scale. The facilities, resources, and vast areas of land available to research and experimental workers are beyond anything I have seen before."[95] The first stop of his tour was the Research Institute of Pomology at Skierniewice, Poland, at the invitation of Szczepan Aleksander Pieniążek. Strawberry production in Poland, Tompsett reported, had boomed from 8,000 to 150,000 tonnes per annum between 1950 and 1966. "The rapid expansion in production," he explained, "has been due to the introduction of virus-free stocks, which have remained free of virus because the winters are too cold for the vector to survive." Likewise, apple production had also benefited from the introduction of frost-resistant rootstock.[96] None of these techniques would have been alien to a British fruit grower. However, their emphasis on the importance of stock-scion relationships and the environment did bear the stamp of Lysenko.

Tompsett's second stop was to the Timiryazev Agricultural Academy near Moscow, where he met V. A. Kolesnikov, head of the Horticultural Department. Astonishing statistics were once again heaped upon the foreign visitor: the area devoted to fruit production in Russia had leapt from 1,625,000 acres in 1917 to 8,750,000 in 1966. Yields of fruit had also risen, thanks to "improved methods of cultivation." Tompsett reported spending much of his time in the rootstock section of the academy, "where the emphasis is on the study of root systems and the effect of the environment on them, linked with the study of the whole tree."[97] In other words, the interaction between stock and scion was still a live subject of inquiry, despite Lysenko's fall from grace. Tompsett moved on to visit more institutions and collective farms in Moldova and Ukraine. In the latter, he was intrigued to see the use of hydroponics to grow apples at the State Farm Technicum in Crimea.[98] He also visited the steppe sub-station near Simferopol, where "large-scale production of clonal rootstocks" occurred.[99] Grafting had apparently allowed orchards to be grown in what were previously considered inhospitable environments. In the North Caucasus, known for dust storms, saline subsoil, and extreme weather, the State Farm Krasnoe covered ten thousand acres of land with orchards, vineyards, and nurseries, thanks to the use of grafted plant stocks "able to withstand heat, drought and frost."[100] Tompsett was impressed by the Soviet Union as a whole. Forests were preserved, parks and roads were immaculately clean, and "conditions of employment are not hard." People had "freedom to change their jobs and to travel."[101] Like many carefully handled Western visitors, Tompsett had apparently been won over by the Soviet Union.

For horticulturalists, belief in the power of grafting did not equate to a belief in graft hybrids. Yet horticulture did provide the necessary conditions for graft hybridization to potentially thrive. At a 1970 lecture Hilary M. Hughes, a regional fruit advisor at the British National Agricultural Advisory Service, noted that "insufficient attention is often paid to the choice of the right combination of rootstock/scion." Amateur gardeners in particular "demand a large tree and they frequently bud [graft] for the amateur trade . . . apple cultivars on strong-growing rootstocks" that invariably proved "entirely unsuitable for garden use." Hughes claimed that this unsuitability stemmed from the amateur gardeners underestimating the extent to which their chosen stock would invigorate their grafted cultivars.[102] Just because a large and strong stock induced vigorous growth in grafted scions and emerging sports (areas of tissue mutation different from the rest of a plant) does not mean anything like graft hybridization had occurred. It does, however, indicate why graft hybridization was a compelling idea among gardeners and horticulturalists who were keenly aware of how grafted plants could influence one another. If such powerful and permanent influences could be seen in real time, it is no great leap to assume that some form of hybridization might have occurred, or that traces of such influence might persist across the generations.

Besides Haldane, other British scientists also began to chafe under Cold War restrictions as the decades wore on. In 1964 John L. Jinks, professor of genetics at the University of Birmingham, argued that "no sooner had chromosomal heredity, with its Mendelian laws of inheritance, been defined and techniques for its recognition developed, than exceptions were described."[103] Such exceptions, according to Jinks, had included the realization that cell organelles could be passed down by plants using heredity mechanisms that did not involve chromosomes.[104] The following year, the British *Times* reported that ongoing experiments from Peter Michaelis, a geneticist at the Max Planck Institute for Breeding Research near Cologne, Germany, had demonstrated "a form of non-Mendelian inheritance" in plants by altering their morphological characteristics without recourse to their nuclei. This, claimed the science correspondent of the *Times*, was only the latest example of an "overshadowed" part of genetics with "a long and respectable history, little related to the extreme position adopted by Lysenko and those in Russia who followed him."[105] Within this wider view of genetics, graft hybrids were entirely possible, although biologists lacked an exact explanation as to why "phenotypic differences" were "graft-transmissible."[106] Support for graft hybridization among breeders, growers, and even ordinary gardeners is, of course, harder to gauge than that of academic scientists. Still, there was a longstanding tradition among British horticulturalists and gardeners that placed great emphasis on the stock-scion relationship. This does not mean that those working in fields and orchards were graft hybridizers. They were, however, open to the idea.

Genetics or Grafting?

Faced with the choice between Western genetics and Soviet grafting, even those with the strongest communist sympathies tended toward the former. By the mid-1950s, when Haldane began his withdrawal from Cold War politics, he had arrived at the conclusion that the potential of graft hybridization had been greatly overstated by Lysenko. Haldane explained the apparent mixing and inheritance of characters following grafting by referring to plant viruses. These viruses could alter the genetics of plants and were known to move across graft junctions. They were therefore a viable explanation for graft hybrid plants that did not undermine genetic orthodoxy. Bateman was never a fan of Lysenko, nor was he particularly inclined to defend the controversial agronomist. He instead saw a future where genetics and dialectical materialism complemented each other. There is no indication that Bateman was opposed to graft hybrids on either scientific or ideological grounds. He did think, however, that the work of Lysenko and his disciples was sloppy.

Their lack of controls and poor data collection meant that even the most compelling examples of graft hybridization beyond the Iron Curtain could not be trusted. Bateman's replication of Felföldy's research reinforced this stance. Graft hybridization could not be taken seriously, as it was simply too rare and too open to experimental errors.

Given this attitude toward graft hybrids from British scientists, why was graft hybridization taken so seriously at British agricultural and horticultural institutions? One explanation is that actors within these institutions were simply more likely to encounter organisms resembling graft hybrids. They may have carried out grafting on a regular basis and were certainly more aware of the horticultural journals and textbooks produced by British graft hybridizers such as Weiss and Neilson Jones. These gave the impression that graft hybridization was still a controversial and active area of biological research prior to Lysenko's interference. A second explanation is that Lysenko and the graft hybrid simply stepped into an area where genetics had failed. Early twentieth-century British breeders had been skeptical about the utility of Mendelian genetics. The respected barley breeder Edwin S. Beaven had declared that "the geneticist will generally offer an explanation of the plant breeder's results after they have been ascertained."[107] This skepticism of scientific experts persisted into the mid-twentieth century.[108] Members of the British Cattle Breeders' Club certainly felt that they had been ignored by academic geneticists, hence their futile efforts to enlist the expertise of Haldane.

Prior to the collapse of the Soviet Union, the careful presentation of a successful horticultural and fruit growing industry lent further credence to Lysenko and graft hybridization. Although Tompsett embarked on his tour of the Soviet Union long after Lysenko had lost his scientific standing and political power, traces of Lysenkoism remained. Vast facilities devoted to the grafting of fruit trees remained intact, as did efforts to expand fruit cultivation into formerly hostile environments. The return of genetics in Russia had not yet made itself felt at the level of these institutes. As one scholar explains it: "Lysenkoism has actually never disappeared from Russian science. Even after 1965, when Lysenko lost his influence, many of his disciples and supporters retained their positions as heads of departments, laboratories and research institutes within the system of agronomic research and higher education. For decades, millions of people studied the basics of biology from textbooks that reproduced Lysenko's ideas."[109] Nor was the persistence of Lysenko—and the graft hybrid—confined to Russia. Within

Western biology, interest in non-Mendelian inheritance continued throughout the Lysenko affair and beyond. All that was needed to resurrect the graft hybrid was a single reliable instance of graft hybridization, stripped of political connotations, to present to a Western audience.

5

Transplantation and Tomatoes

⟨decorative flourish⟩

IN AUGUST 1957 TWO BRITISH BIOLOGISTS FOUND THEMSELVES FACE-TO-face with Trofim Lysenko. Donald Michie, a geneticist and committed Marxist, had traveled to the Moscow Institute of Genetics to meet Khilia F. Kushner, a Lysenkoist who bred hybrid poultry. Upon his arrival, Michie found that Kushner was not yet available. Lysenko, however, was. The controversial agronomist had lost much of his previous status, including the directorship of the V. I. Lenin All-Union Academy of Agricultural Sciences. Still, he had kept his post as director of the Moscow Institute of Genetics and was quite prepared to defend his biology. A sympathetic Michie entered the interview as a believer in the existence of graft hybrids. He asked Lysenko about the possibility of exchanging experimental data on grafting between laboratories in the East and West, explaining how the disagreement between L. J. M. Felföldy and Angus John Bateman had unfolded. Lysenko was dubious, claiming that graft hybrids were so easy to make that anyone could already do so if they wished. "I challenged him to describe such an experiment," wrote Michie. "He at once suggested one in barley, which he claimed could be done in five minutes." Lysenko claimed that grafting the embryo from an awned variety of barley to the endosperm of an awnless variety would result in both awned and awnless offspring.[1]

Also present at the Lysenko interview was Michie's wife, the Oxford-trained zoologist Anne McLaren. McLaren was a powerful force in Cold War biology. Later in her scientific career she would become the first female officer of the Royal Society, and an authority on bioethics. She was a Communist Party member whose research interests focused on mice chimeras.[2] She was also a steadfast believer in the existence of graft hybrids. "I belong to the generation of biologists," recalled McLaren of her early career, "that finally got fed up of the elegance and mathematical exactitude of Mendelism." She agreed with fellow zoologist Avrion Mitchison, who believed that the "stimulus" of the Second World War was partly responsible for Western biologists examining instances of heredity "by means other than the traditional chromosomal route."[3] For McLaren, one of these routes was graft hybridization. Her interpretation of Marxism gave her a clear ideological incentive to argue for the existence of graft hybrids. Yet as we have seen in the cases of J. B. S. Haldane and Bateman, Marxist biologists in the West were often skeptical of graft hybrids, or denied their existence outright. McLaren would complain that the scientific work on graft hybridization was "bedevilled at the time by being all mixed up with the politics of Lysenkoism."[4] She would spend the 1950s largely avoiding overt references to Cold War politics, instead attempting to place the graft hybrid firmly within the realms of respectable experimental biology.

McLaren employed different strategies to bring the graft hybrid across international barriers. The first was through covert efforts, helping a Czech biologist named Milan Hašek to publish his experiments creating hybrid poultry via embryonic blood transfusion in Western journals in the 1950s. The second involved replication. Shortly after meeting Lysenko, McLaren traveled to Yugoslavia to learn grafting techniques from botanist Ružicka Glavinic. She then attempted to grow her own graft hybrid tomatoes in London, although she was ultimately unsuccessful. McLaren also attempted to acquire experimental data on plant grafting from communist China. Finally, in the late twentieth and early twenty-first centuries, McLaren attempted to ally graft hybridization with new biological sciences and techniques. She used epigenetics to support the inheritance of acquired characteristics, collaborated with contemporary plant breeders, and connected graft hybridization with modern forms of biotechnology through genetic engineering.

Two Cold War Interpretations of Transplantation

In 1952 McLaren acquired funding from the Agricultural Research Council and found herself working alongside Peter Medawar, a British biologist based at University College London. Medawar had spent his career working on immunology and transplantation. His interest in the subject originated during the Second World War, when a doctor approached him and asked if it might be possible to treat a severely burned airman using skin grafts grown from cells.[5] Sadly, such a treatment was not possible, but Medawar was intrigued by the idea. In 1949 he turned his attention to the "freemartin" problem in cattle, where nonidentical twin cattle, one male and one female, are born with the female calf sterile. This problem was both a scientific puzzle and a commercial setback for cattle breeders who wanted fertile cows. Hugh Paterson Donald, an Edinburgh veterinarian, suggested that skin grafts might be used to distinguish these freemartins from identical twins. Donald reasoned that the nonidentical twin would not accept grafts from his or her genetically distinct sibling or vice versa. Their immune system would identify the graft as "alien" and destroy it. Yet when Medawar put this theory into practice, he found that the genetic orthodoxy of the time did not stand up to scrutiny. Nonidentical cattle twins accepted skin grafts well into adulthood, which suggested that tolerance had less to do with genetics than with cross-circulation of blood in the womb. This phenomenon was Medawar's first piece of evidence that a kind of immunological "tolerance" could be acquired by an organism over time.[6] Backed by a team at University College London, Medawar spent the early 1950s replicating his cattle experiments in other animals. He found that embryonic chickens and mice who received donor cells also gained the ability to accept transplants from that same donor in later life. In a 1953 paper in *Nature*, Medawar and his team introduced the concept of "actively acquired tolerance." This idea was the opposite of the more familiar acquired immunity, where vaccination, infection, or exposure to antibodies allows the immune system to "memorize" and respond to microbiological intruders. Medawar found that grafting foreign tissue onto embryonic mice did not incur this response. The bodies of these mice accepted the graft and were more tolerant of grafts from the same source in the future. They had effectively "acquired" tolerance, wrote Medawar, as "resistance to a graft transplanted on some later occasion, so far from being heightened, is abolished or at least reduced." His team claimed to have found a solution—albeit only a "laboratory solution"—to the medical barrier that Medawar had run up against during the Second World War: "the problem of how to make tissue homografts immunologically acceptable to hosts which would normally react against them."[7]

At this point, McLaren threw a proverbial spanner in the works. In 1953 she and Michie attended the International Congress of Microbiology in Rome, where they first heard about the young Czech scientist Milan Hašek. Jaroslav Sterzl, a microbiologist in the Department of Experimental Biology and Genetics at the Biological Institute of the Czechoslovak Academy of Sciences, informed McLaren that Hašek had found some exciting results during his work on parabiosis—when two organisms are surgically attached and share physiological systems, such as the circulatory system—in chicks. On a superficial level, these results were very similar to those obtained by Medawar and his London team. Once Hašek separated his embryonic chicks, he had found that the adult chickens could freely accept grafts from their former partners without eliciting an immunological response. He had even connected the circulatory systems of different species. Hašek found that when the blood of a turkey and a chicken were mixed, evidence of this connection in the form of foreign blood cells remained in the turkey for as long as eight weeks following their separation. He concluded that their presence could only be explained by the bird creating foreign blood cells in its own circulatory systems. According to immunologist Leslie Brent, Medawar's student and coauthor, this finding "has never been adequately explained."[8] It is no coincidence that this ability to swap tissues between different species sounds reminiscent of graft hybridization. Hašek claimed that he had demonstrated the validity of Lysenko's biology in animals by creating hybrids through parabiosis. The Czech biologist was operating under a very different scientific and political regime than Medawar was. Following the Communist coup of 1948, Lysenko's biology swept across Czechoslovakia. In Brno, the hometown of Gregor Mendel, the geneticist Jaroslav Kříženecký found his subject banned and was driven from the university. In 1958 he was thrown in prison for daring to compare Lysenko's rejection of genetics with those who had once rejected gravitational laws. His former student Vítězslav Orel was demoted on the grounds that he was an "individualist."[9] After the Second World War, Hašek had studied at Charles University under Bohumil Sekla, another advocate of Mendelian genetics. Unlike many Mendelians, however, Hašek had joined the Communist Party in 1950. Party membership brought scientific advantages. He was soon offered a prestigious position at the Czechoslovak Academy of Sciences, with the Communist regime prioritizing projects that supported Lysenko's doctrine.[10] Here Hašek sought to toe the official party line and gain support for research, stating that Lysenko had disproved Western genetics and arguing that parabiosis in animals provided further support for the existence of graft hybrids.

Following the 1953 congress in Rome, McLaren returned to London and relayed her findings on Hašek to a "somewhat disconcerted" Medawar. She related that his discomfort was due not to political ideology but to the sudden fear that his research on immunological tolerance might have been "scooped" by an unknown rival.[11] However, his fears soon proved to be unfounded. All the experiments conducted by Hašek were effectively identical to those previously conducted by Medawar and his team. Just as importantly, Hašek's interpretation of his experimental findings through a Lysenkoist lens was completely unacceptable to Western biologists. The stage had been set for what could have been the latest ideologically driven battle of Cold War science. On one side were Medawar and the theory of acquired immunological tolerance. On the other side of the Iron Curtain were Hašek and his interpretation of the same results as graft hybridization. A clash between the two opposing ideologies never came. Hašek began to alter his interpretation of parabiosis as he became increasingly aware of the Western scientific literature on immunology. When Medawar and his colleagues published their 1953 paper in *Nature*, Hašek changed his language to match theirs. When he next summarized his experiments on parabiosis and skin grafting in *Československá biologie* (Czechoslovak biology), Hašek abandoned all his discussion of genetics in favor of immunology. He also abandoned Lysenko's "vegetative hybridization" in favor of "vegetative rapprochement," bringing his scientific language away from graft hybridization and toward immunological tolerance.[12] The story of Medawar and Hašek's Cold War science might have been an uplifting tale of scientific compromise and collegiality across political divides. Instead, two very different interpretations of these events followed. What we could call the "standard" account was given by Brent. Brent had arrived in Britain on the Kindertransport just prior to the outbreak of the Second World War. He fought with the British Army in Italy and on the Rhine, studying zoology at the University of Birmingham after the war. He was taught by Medawar, who later encouraged Brent to follow him to University College London and pursue a doctorate. It was Brent who first introduced their work on grafting in mice to amazed delegates of the Society for Experimental Biology in 1953.[13] Brent recalled that Medawar reached out to Hašek shortly after McLaren's 1953 report. He and Medawar met Hašek in person in 1954 and attempted to persuade him—if any further persuasion was needed—that many of his findings could instead be interpreted through their own theory of acquired immunological tolerance. Their efforts, according to Brent, were successful.[14] Although Hašek could not denounce Lysenko, he built a close relationship with Medawar and began to receive the latest literature on transplantation and immunology from the West.

Medawar eventually helped Hašek publish his findings for an English-speaking audience in *Proceedings of the Royal Society* in 1956. This time, however, all mentions of Lysenko and graft hybrids were safely removed. "Embryonic parabiosis," Hašek now argued, "brings about an immunological tolerance which persists for a long time, sometimes perhaps throughout the individual's life."[15] In the same year, Medawar and his team published more details on their own grafting experiments. By this time they had used grafting to induce tolerance in mice, rats, rabbits, and birds. The images that Brent had used to wow members of the Society for Experimental Biology were now in print. They included white mice with engrafted patches of black fur, white chickens with patches of brown feathers, and even a duck with a graft of chicken skin.[16] "Tolerance," wrote Medawar and the University College London team, referred specifically to the "weakening or suppression of reactivity caused by the exposure of animals to antigenic stimuli before the maturation of the faculty of immunological response."[17] Acquired tolerance was temporary, was limited to transplants between similar species, and did not involve any genetic exchange. Medawar described his animal subjects as chimeras, with one exception. Referring to his work with freemartin cattle, he noted that tolerance "was long enduring, and some of the subjects of these earlier experiments are still demonstrably 'graft-hybrids' to-day."[18] Although his choice of words is interesting, the quotation marks imply that his description of the cattle as graft hybrids was not meant to be taken literally.

The second interpretation of Medawar and Hašek's Cold War collaboration came from McLaren. During the early 1950s, McLaren was engaged with Marxist biology and was particularly fascinated by graft hybrids. Her introduction of Hašek to Medawar had been an early attempt to bring graft hybrids back to the attention of Western biologists. In 1954 she and Michie visited Hungary, where they met with L. J. M. Felföldy at the Biological Research Institute in Tihany and reported his successes in "vegetative hybridization." They also met with a Professor Györffy, director of the Hungarian Academy of Science's Institute of Agrobiology in Budapest. Györffy had been attempting to replicate the work of the French botanist Lucien Louis Daniel and leading Lysenkoist Ivan Yevdokimovich Glushchenko by grafting different tobacco varieties together. McLaren and Michie also visited the Institute of Small Animal Husbandry at Gödöllő. Here they encountered a series of experiments very similar to Hašek's. They met a Dr. Mészáros, who claimed that transfusing blood from turkeys to chickens led to offspring that were larger and faster growing. He had also attempted to use blood transfusion to create sheep-goat hybrids, citing Ivan Vladimirovich Michurin's theory of "vegetative approximation."[19] A year after their trip, in 1955, McLaren and Michie wrote an essay that appeared alongside Haldane's "Some Alternatives to Sex" article, in which they praised the work of "the Russian plant hybridizer Michurin."[20]

McLaren continued to view Hašek's findings through a Marxist lens. Decades after the original incident, she insisted on interpreting Hašek's results in much the same way that he had originally done. In her view, Hašek's turkey-chicken chimeras were a "type of vegetative hybrid," with embryonic parabiosis effectively "analogous to combining them [the chickens and turkeys] sexually by crossing." McLaren's interpretation did account for why Hašek did not attempt to breed his poultry to check for signs of hybridization persisting beyond a single generation. All that was needed for proof of graft hybridization, at least according to McLaren, was signs of heterosis—or hybrid vigor (fig. 10)—in plants or animals that had received grafts. In her own research, McLaren had predicted that hybrid vigor would be observed in mice chimeras, or, as she termed them, "another type of vegetative hybrid."[21] This is a key distinction, which sets McLaren apart from the other graft hybridizers we have met thus far. To make a graft hybrid, all one had to do was induce a long-term physical effect in an organism by using transplantation or transfusion. This change did not necessarily need to be passed down the generations for the graft hybridizer to claim success.

McLaren was also able to offer other insights into the motivation and goals of Hašek's parabiosis experiments. She recalled that she and Michie had visited Hašek in 1956, by which time the "Lysenkoist atmosphere" was "already fading." McLaren claimed that Hašek was more heavily influenced by Michurin than by Lysenko. Hence, he "was not concerned with transmission of characteristics to the next generation (he never mated his experimental animals)," but instead pursued parabiosis to investigate "assimilation" over hybridization. The resulting chicks had, however, shown signs of hybrid vigor. McLaren claimed to have already seen this phenomenon in her mice chimeras and argued that it "is well known in plant graft hybrids." On Hašek's apparent shift toward immunology, McLaren suggested that "he changed his terminology (and in accordance with the times, his political—not scientific—position) rather than his mind." It was only when Hašek began writing in English that he shifted toward the term *immunological tolerance*, a result of his newfound British connections and access to "more up-to-date literature."[22] Can we take McLaren's interpretation seriously, or was it driven by her

FIGURE 10. Culture comparison of grafted and nongrafted plants. Part of the economic appeal of hybrid plants and animals is heterosis, or hybrid vigor. In addition to attempting to find heritable traits in grafted organisms, British zoologist Anne McLaren argued that signs of heterosis following transplantation was evidence of graft hybridization. Lucien Louis Daniel, *Les mystères de l'hérédité symbiotique: Points névralgiques scientifiques, pensées, théories et faits biologiques* (Rennes: Roger Gobled, 1940), plate II.

politics? Hašek's political leanings certainly shifted against communism over time, although he professed loyalty to the party into the 1960s. This display of allegiance drew in more funding for his research group, which became an institute at the Academy of Sciences in Czechoslovakia.[23] These practical motivations make it difficult to draw any firm conclusion on Hašek's real beliefs.

Between these different accounts of Medawar and Hašek's exchange emerges a more nuanced interpretation of transplantation and acquired tolerance. It is not necessary to portray the different Cold War views on acquired tolerance as politically incompatible. McLaren would recall that during the 1950s she "was culturing mouse embryos in DNA prepared for [her] by Medawar to see if it would have any effect on development—but not surprisingly it didn't!" Medawar, it seems, was quite open to new possibilities in genetics. "It was a time," recollected McLaren, "when the idea of infectious heredity was very fashionable." McLaren explained that University College London was not immune to this new trend in biology: Medawar and his postdoctoral researcher, Rupert E. Billingham, "thought that they had discovered infectious transmission of pigment from cell to cell in guinea pigs. They grafted skin from black areas of piebald guinea pigs onto white areas, and the black areas spread outward—but alas, it turned out to be due just to black cells migrating away from the grafts and not infectious transmission of anything."[24] This idea was essentially identical to the one Haldane invoked in defense of the graft hybrid. It explained how hereditary characteristics could be exchanged between grafted plants and animals using a relatively noncontroversial mechanism. McLaren was also interested in accounts of hereditary characteristics passed on through infection. On August 4, 1960, she received a letter from Herbert Butler Parry, a veterinary surgeon at the University of Oxford. Parry's main interest was scrapie, a degenerative disease found in sheep and goats. He informed McLaren that the disease was both inherited and could be passed artificially between different sheep and goats. "The fact that Scrapie appears to be behaving as a mendelian recessive," he told McLaren, "and at the same time transmissible by inoculation, intrigues us very much indeed."[25]

The Cold War clash over the nature of heredity and the role of blood in inheritance left other legacies. Medawar went on to share the 1960 Nobel Prize in Physiology or Medicine for the discovery of acquired immunological tolerance. Yet he may have felt some measure of disquiet about his censorship of Hašek's original interpretation. In 1964 Medawar provocatively asked,

"Is the scientific paper a fraud?" in a BBC radio broadcast, criticizing the inductive and artificial structure of the typical scientific paper. Scientists, declared Medawar, are ordinary human beings, guided by imagination, intuition, and personal philosophy. By ignoring these elements, he claimed, we neglect a fundamental part of the scientific enterprise.[26] Medawar's general aversion to the unnatural structure of scientific papers can be adequately explained through his admiration for the philosophy of Karl Popper, whom Medawar had personally invited to Cambridge while Medawar was still a student.[27] Looking beyond this interest in Popper, however, we can speculate that Medawar perhaps felt some responsibility for his own role in the removal of Hašek's personal philosophy. Although Medawar's actions had allowed Hašek to be published in the West, by doing so he had inadvertently helped to maintain a system he philosophically disagreed with.

McLaren's approach to the Hašek affair was an example of what we might term—in the spirit of the Cold War—*covert science*. A widespread interest in the possibility of the exchange of genes, whether through infection or the exposure of embryos to foreign DNA, indicated that there was some hope that her colleagues in the West could accept the existence of graft hybrids. Within the politically charged atmosphere of the time, though, acquiring trustworthy information on grafting beyond the Iron Curtain was not straightforward. McLaren therefore embarked on what would be the first in a series of personal campaigns designed to bring reputable research from Eastern Europe to the attention of her colleagues. Hašek's acceptance in the West had, however, come at the cost of him abandoning his initial Michurinist interpretation of parabiosis. Despite some unexplained aspects to the Hašek experiments, a scientific consensus emerged that Hašek's poultry were chimeras rather than graft or vegetative hybrids. If McLaren wished to reinvigorate graft hybridization in the West, she would have to provide a trustworthy and irrefutable demonstration. The best way to do this would be to create graft hybrids herself.

Plant Grafting in Yugoslavia and China

Just prior to her relocation to the Institute of Animal Genetics at Edinburgh University, McLaren once more attempted to carry the findings of Lysenko-supporting biologists in Eastern Europe across the Iron Curtain. If graft hybridization in animals could not convince her Western contemporaries of the virtues of Marxist biology, McLaren thought, then perhaps she could create plant graft hybrids herself. An opportunity began to take shape in the late 1950s, when she heard of some intriguing tomatoes created by Glavinic, based at the Faculty of Silviculture at Belgrade University, Yugoslavia. On January 1, 1958, McLaren described how Glavinic had invited her to Belgrade to repeat the tomato grafting experiments there. "I am attracted by the idea," wrote McLaren, "as although her results are unsatisfactory in various ways, the claims are very striking, and she herself impressed me as an honest and sincere person, though without much general training." There were practical considerations, too. "Yugoslavia is both politically and geographically more accessible than the USSR," mused McLaren, "also Glavinic's results are actually more striking than those of Glushchenko."[28]

In Yugoslavia, Lysenko's biology was never officially endorsed or denounced by the ruling Communist Party. Lysenko's supporters did not dominate biology or agriculture, although they did gain a following among Yugoslavian agronomists. One of these agronomists was Glavinic, who lent her support to "Michurinist science" in the late 1940s, pointing to graft hybridization as a means of implementing the new biology.[29] References to Lysenko in academic journals were not numerous to begin with, effectively withering away by the mid-1950s. The only aspect of Lysenko's biology that survived to the end of the decade in Yugoslavia was graft hybridization. The theoretical importance of the technique, combined with the occasional glimpse of an apparent success, meant that graft hybrids were still considered a viable avenue of research after most of Lysenko's ideas had fallen out of fashion.[30] For McLaren, Yugoslavia may well have represented a kind of "neutral zone," where graft hybrids could be investigated without undue political or ideological pressure from either side of the Cold War. The fact that some agronomists continued to be interested in graft hybrids, independent of Lysenko, must also have offered her some grounds for hope.

The British Agricultural Research Council was open to the idea of McLaren replicating graft hybridization in Yugoslavia. On February 21, 1958, McLaren received a letter from her funder giving her project the green light, provided she obtain some expert insights on plant grafting before her departure. The East Malling Research Station in Kent and the John Innes Horticultural Institution in Cambridge both had "people who may have had some experience in this field."[31] In a description of the planned project for University College London, titled "A Proposed Experiment on Graft Hybridisation in Tomatoes," McLaren explained her intentions. She began by describing contemporary scientific interest in the "transmission of hereditary characters by non-sexual means," notably by bacteria and viruses.

"A similar phenomenon has been claimed to occur in higher plants," wrote McLaren, "and has been termed 'graft hybridisation.'" For the purposes of her Yugoslavian project, she explained that a graft hybrid emerged if "a plant of one variety or species is grafted onto another, and the progeny (or 'first seed generation') of one component of the graft develops characteristics specific to the other component."[32] This traditional concept of a graft hybrid differed from the definition McLaren adopted after the Hašek encounter, when she declared that any organism that showed signs of hybrid vigor following grafting could be categorized as a graft hybrid. McLaren may have adopted this latter definition later, or more likely was shifting her definition of graft hybridization to suit different contexts.

A review of scientific thinking on graft hybridization must have given McLaren some confidence. Charles Darwin, of course, had discussed graft hybrids, albeit based on "anecdotal evidence." In 1946 P. S. Hudson and R. H. Richens, of the Imperial Bureau of Plant Breeding and Genetics, had "concluded that the induction of heritable modifications by grafting is a possibility, but at that time undemonstrated."[33] McLaren had also kept up with accounts of graft hybridization from the Soviet Union, although she noted that no genetic analysis of these plants had been carried out after grafting. She thought that prior attempts at replicating graft hybrid experiments in the West were inadequate. Fellow Marxist Bateman had tried and failed to replicate the tomato-grafting experiments of Felföldy. The Hungarian botanist's later withdrawal of his claims, declared McLaren, "leaves many more questions unanswered, serving only to increase the confusion which surrounds this subject."[34] The only other attempt at graft hybridization in the West that McLaren was aware of was made by Leo Sachs, a German-born Israeli whose family had fled from Germany to England after the rise of the Nazis. Sachs studied agriculture at Bangor University, Wales, graduating in 1948 before entering Trinity College in Cambridge to study plant genetics in 1951.[35] Unlike most students, Sachs decided to spend the gap between his academic programs replicating Lysenkoist work on tomato grafting. Under the tutelage of esteemed British plant breeder G. D. H. Bell, Sachs found that a strange leaf shape that appeared in the grafting experiments of the Russian botanist Glushchenko was also appearing in nongrafted tomato plants. "The results of this experiment," concluded Sachs, "clearly show that in the year of grafting there has been no observable influence on fruit colour or leaf-shape by either stock or scion. These results, therefore, do not support the claims of the Russian workers."[36]

McLaren pressed ahead with her plan to visit Yugoslavia, arguing that since Sachs had not used the same tomato varieties the Russian Lysenkoists had, his "lack of positive results is of little evidential value." She was open to the idea, shared by Haldane, that graft hybrids could be the result of plant viruses carrying genetic material across the graft junction. This "might well be true," wrote McLaren, "but it would not lessen the importance of the phenomenon."[37] There were, of course, other less palatable explanations for Glavinic's results. McLaren had already dismissed the idea of scientific fraud as unlikely, appraising Glavinic as "honest" and "sincere."[38] She was also convinced that Glavinic's experimental technique was sound. In a plan of her proposed visit to Yugoslavia, McLaren described how Glavinic had grafted the two tomato varieties Potato Leaf and Golden Trophy. Both varieties had been self-fertilized for four generations prior to their grafting in 1950, and both possessed characteristics that followed Mendelian laws. Unlike other Lysenkoists, Glavinic had used experimental controls, growing a crop of forty Golden Trophy plants that "continued to breed true" across the generations. According to Mendelian genetics, the seeds of Glavinic's grafted plants should also have resembled the Golden Trophy variety. They instead took on a hybrid appearance. McLaren declared that it was "impossible to account for the results in terms of accidental cross-pollination between the varieties." Unlike Felföldy, Glavinic and her graft hybrids could not be dismissed by Western biologists on the grounds of "poor experimental technique."[39]

In May 1958 McLaren was in Belgrade, where she worked alongside Glavinic. The pair planned to produce four sets of grafted tomato plants. Two of these sets would be raised in Belgrade by Glavinic and two would be taken to London by McLaren. Some of McLaren's plants did not survive the trip—they were packed too tightly and rotted en route. The surviving plants found a home in the greenhouses at Nuffield Lodge in London. To obtain this space, McLaren had contacted Dan Lewis, the pomologist who had tasked Bateman with replicating Hungarian graft hybrid experiments in 1949, now a professor of botany at University College London. This was Lewis's second encounter with graft hybrids. "I would be very interested to hear the results of your experiments," he wrote to McLaren, "although I had personally put the matter out of my mind after I got Bateman to repeat the work that Felfoldy had done previously, but it may be that some other factor is involved, perhaps a virus."[40]

Once safely planted in the greenhouse, some of McLaren's tomato plants began to flower. Only one produced fruit. A single red tomato was taken from the scion of a plant that usually produced yellow tomatoes. In February 1959 McLaren wrote to Lewis again, seeking help with the low fruit count of her grafted plants and describing her unusual tomato. Lewis guessed that the tomato varieties McLaren had brought back from Yugoslavia were not suited to the British climate, or that the soil they were grown in contained too much nitrogen. On the plus side, the single fruit that McLaren had managed to produce was certainly unexpected. Lewis suspected that McLaren had produced a chimera, writing: "The one red fruit on the yellow fruited plant is very interesting, if at this stage, somewhat obscure. I have never heard of such a change on normal, ungrafted plants at the end of the season or at any other time. You may have a chimaera, which are very readily produced within the tomato family, and if this is so, your seed next year may or may not produce yellow fruited plants, according to how the two tissues are mixed within the scion."[41] Obtaining a chimera was no small feat. After all, in the late nineteenth century, Darwin himself had tried and failed to do so. The German botanist Hans Winkler had obtained only a handful of chimeras after grafting thousands of plants in the early twentieth century. While her chimera was technically impressive, however, McLaren had wanted more. Her aim had been to replicate the strange tomatoes originally described by Glavinic under controlled conditions. With this goal now in peril, McLaren wrote to Glavinic for an update on the grafted plants in Belgrade. The latter informed McLaren that she had left the Faculty of Silviculture, "so that in general I do not know if there were any fruits on the grafts + if they ripened." Glavinic encouraged McLaren to revisit Belgrade and continue grafting plants in London. "I believe," she wrote, "that you will obtain vegetative hybrids."[42]

It was not to be. After making inquiries with Zora Lazarevic, another member of the Silviculture Faculty in Belgrade, McLaren found that the Yugoslavian tomatoes had not produced any fruit. They were of no further use in her experiment. How, then, to explain the appearance of the hybrid tomatoes that had drawn her to Belgrade in the first place? McLaren concluded that "during the 1957 growing season in Belgrade, the bagging of ZT [referring to the grafted plants] flowers was inadequate."[43] The tomatoes obtained by Glavinic were probably conventional hybrids, formed when a stray insect carried pollen from one tomato variety to another. This explanation for strange fruits formed after grafting had also emerged when Bateman had tried and failed to replicate experiments first conducted by Felföldy in Hungary. When it came to proving the existence of graft hybrids, tomatoes turned out to be a less-than-ideal experimental organism. Pollen was carried between different varieties by tiny insects, which seemed to effortlessly evade the best efforts of Eastern European horticulturalists and botanists to keep them out.

All was not lost for McLaren, as the Lysenkoist fascination with graft hybrids was a phenomenon not confined to Eastern Europe. In China, the introduction of "Michurinist biology" reflected Sino-Soviet relations, with pro-Michurin textbooks and articles translated for a Chinese audience by the Sino-Soviet Friendship Association and All-Union Society for Cultural Relations with Foreign Countries.[44] By 1952 the Chinese Communist Party had banned "Mendelist-Morganist" views, although scientists were not required to profess their support for Michurinism.[45] Some biologists were caught up with the promise of the new Soviet biology. Among them were Teh-Ming Tsu and his colleague Yu-Seng Chao, members of the North China Agricultural Research Institute. In a 1955 paper Tsu explained that he had "unexpectedly obtained a short piece of literature regarding Michurinism while working in the border region of Shansi, Charhar and Hopei Provinces as early as 1944." Two aspects of Michurin's work intrigued Tsu. The first was the Russian horticulturalist's "extraordinary success" in breeding fruit trees. The second was Michurinism's rejection of Western genetics and formulation of a new theory of heredity. For Tsu, who claimed to once be "deeply imbued with the Mendel-Morganian genetics," the Soviet dismissal of the gene "led him to consider anew what the hereditary phenomena of animals and plants ultimately amount to after all."[46]

Tsu was most struck by Soviet claims that new plant varieties could be created by grafting. When he arrived in Peking (now Beijing) in the wake of the Communist armies in 1949, he began his own grafting experiments. They were intermittent and often interrupted, conducted under "war conditions" and often beset by heavy rainfall that drowned the plants.[47] Tsu chose to graft two different varieties of eggplant, while Chao worked with tomatoes. For his eggplant grafting experiments, Tsu carefully selected varieties that were well known and had a low mutation rate. One was the nine-leaf eggplant, with round purple fruits. The other was the white eggplant, which bore elongated green fruits that gradually turned white, then yellow. From 1951, Tsu began grafting branches of white eggplants onto stocks of the nine-leaf variety. He knew what ratios of hybrid offspring were predicted by Mendelian laws and was aware of the dangers of accidental cross-pollination. "For the purpose of avoiding accidental pollination on the experimental plants," he reported, "the authors wrapped now and then the flower buds with paper bags or gauze covers, and tied them to inhibit blossoming."[48]

Although Tsu was not able to successfully cultivate many plants, he did manage to obtain some fruit from his grafted eggplants. He observed no change in their expected color, but did notice that the fruits from the white eggplant scions were ovoid or rounded, when they were usually elongated. Tsu took this as evidence of "graft variation beyond question." He was also able to obtain seeds from these fruits and plant them to grow a second generation. These plants were purple and had purple fruit, suggesting that the color characteristic of the purple nine-leaf eggplant had crossed the graft junction and hybridized with the white eggplant.[49] He repeated his experiment again in 1952. This time, heavy rainfall conspired with a frost to rob Tsu of all but two of his plants. Once again, these survivors bore rounded purple fruit. Their seeds resulted in plants that had white or purple fruits. Tsu obtained similar results when he grafted the nine-leaf eggplant onto a stock of white eggplant. He had also attempted cross-species grafting, transplanting the scion of the nine-leaf eggplant to a stock of Illinois Pride tomatoes in 1949 (an example of a similar plant is depicted in fig. 11). There were initially no signs of hybridization, until Tsu planted seeds obtained from his eggplant-tomato grafts in 1951. Then he observed a whole range of strange variations, with plants of different sizes and leaf shape. Odd fruits also emerged, with eggplants of red and purple, "which looked quite pretty in appearance." Tsu concluded that "grafting between widely different species may serve as a way to the creation of new species."[50]

Summarizing his results, Tsu announced his support for "the principle of vegetative hybridization as discovered by Michurin and further developed by Lysenko." One of the most significant aspects of his experiments, he announced, was to break through the usual barriers that prevented the hybridization of different species. Tsu had tried and failed to obtain tomato-eggplants through cross-pollination. Grafting, however, seemed to have worked. He believed that the variations in fruit color, size, and leaf shape in his grafted tomato-eggplants could become "hereditary" and "act as stocks for breeding new varieties of eggplant."[51] Obtaining graft hybrids was apparently quite straightforward. For Tsu, the real challenge facing botanists and horticulturalists was in predicting and explaining what changes would be induced by grafting different species. He was also concerned with how to control the inheritance of different characters, whether obtained through grafting or mutation.[52] In other words, a kind of nongenomic equivalent for

FIGURE 11. Eggplant (aubergine) grafted onto tomato. Shortly after the Chinese Civil
War, Teh-Ming Tsu of the North China Agricultural Research Institute began grafting
varieties of eggplant together, as well as grafting eggplants onto tomato plants. The strange
appearance of the resulting fruit convinced him that Michurin and Lysenko were right.
Graft hybridization was real and could be used to create new species. Lucien Louis Daniel,
*Les mystères de l'hérédité symbiotique: Points névralgiques scientifiques, pensées, théories et faits
biologiques* (Rennes: Roger Gobled, 1940), plate XXV.

Mendelian genetics that could predict the outcomes of particular crosses was required to bring graft hybridization from the experimental greenhouse to Chinese horticulture.

By the mid-1950s, however, Lysenkoism was on the defensive in China. A blow against graft hybridization was struck by the influence of German agronomist and plant breeder Hans Stubbe. He had repeated some of Lysenko's vegetative hybridization experiments and found no evidence that grafting could produce hybrids. Stubbe visited China in 1955 to present his research at Beijing Agricultural University. His criticism of Lysenko—whom he had met in person in the Soviet Union—reached the Central Committee of the Chinese Communist Party and Chairman Mao Zedong.[53] The downfall of Lysenko in the Soviet Union was followed by outspoken criticisms in China. At the 1956 Qingdao Genetics Symposium, a new generation of Chinese geneticists trained in the United States were able to voice their frustration with the monopoly maintained by the Lysenkoists in biology.[54] Geneticists emerged relatively unscathed from antirightist campaigns and the Great Leap Forward, while the Cultural Revolution targeted scientists on both sides of the genetics debate.[55] McLaren encountered this new reality when she wrote to Tsu in April 1958, hoping to obtain more information on his graft hybrid experiments from January of that year. Tsu supplied a copy of his article, but informed McLaren that his eggplant experiments had only just restarted. "The experiment is now just being regained for some further observation," wrote Tsu, "and I am very glad to send you any results obtained for your reference."[56]

Tsu did not continue his grafting experiments, or at least never got back to McLaren about them. China and Yugoslavia had both presented McLaren with the promise of seeing plant graft hybridization in action. Yet these transnational exchanges had not provided her with the desired results. Why? Part of the blame must lie with graft hybrids themselves. Despite the considerable interest and influence they represented for scientists across the twentieth century, graft hybrids were still extraordinarily difficult to make. Extensive time and investment were necessary requirements to achieve a possible case of graft hybridization. Even then, there were no guarantees. One might end up with an organism that could be written off as a chimera or, worse still, no hybridlike organism at all. With this in mind, we can also consider McLaren a victim of timing. Her efforts to locate and reproduce graft hybrids came in the late 1950s, when Lysenko and his approved techniques were in retreat. Classical genetics was on the rise in both China and Yugoslavia, meaning that McLaren did not have a vast range of options in terms of botanists or agronomists to approach. Fewer graft hybridizers meant less graft hybridization, leaving McLaren with a limited range of examples to draw upon.

Mendel and Michurin Today

Throughout the 1950s, McLaren had been cautious about being drawn into the Lysenko controversy. She reserved her praise for Michurin, the Russian horticulturalist whose ideas on graft hybridization had later been adopted by Lysenko.[57] In 1957, however, McLaren traveled with Michie to Moscow. Here they met Lysenko in person, with Michie conducting an interview with the sidelined agronomist.[58] The interview was a strange one, not least because Lysenko apparently had less faith in the utility of graft hybrids than his Western guests did. When Michie raised the topic, Lysenko declared, "I no longer believe that grafts between different species can give rise to true vegetative hybrids. You can only get chimaeras—a mixture of cellular elements from the two plants with no specific influence of one upon the other." Whether stung by political setbacks or scientific attacks, Lysenko now claimed that only grafting between different varieties of the same species could result in genetic changes. When Michie pointed out that Michurin had created hybrids by grafting apple and pear trees, Lysenko suggested that cross-species grafting only served to create a "general destabilising effect on heredity . . . making the genotype more accessible to environmental training."[59]

A dissatisfied Michie was ready to drop the subject when a fruit breeder from Leningrad named Tetyev entered the room. Unaware of the earlier conversation, Tetyev began describing how he had successfully grafted gooseberries onto blackcurrants, the gooseberries subsequently acquiring resistance to blight. This trait was passed down the generations when they were asexually propagated. Michie challenged Lysenko to explain this contradiction, given that the agronomist had just declared it impossible to create graft hybrids by crossing different species. "Tetyev's work is of practical rather than theoretical importance," responded Lysenko. "The gooseberry is a chimaera. There is no physiological approximation between graft and stock." Michie again pushed him on how it was possible that blight resistance had been transferred to gooseberries following the graft. Lysenko gave his audience a somewhat mangled lesson in what was essentially the Western view of the chimera-graft hybrid distinction: "Taking a branch of gooseberry from Tetyev, he held it motionless and stared intently at it. 'This,' he said after a long pause, 'is a gooseberry. Everyone would say it was a gooseberry. But if you raise a sexual generation from it, alongside the vast majority of its progeny, which will also be gooseberries, it will throw off a small proportion of blackcurrents [sic], and a rather larger proportion of gooseberry-blackcurrent [sic] hybrids. So you see blackcurrent [sic] elements must be incorporated into the gooseberry germinal tissue, but not truly combined, since their products emerge uncontaminated in the seed generation.'"[60]

Shortly after this interview, McLaren began her investigations into plant grafting in China and Yugoslavia. She focused on crossing different plant varieties through grafting. During this period, McLaren reportedly felt "irritation at the neglect of the role of environmental influences" in biology, and in 1959 moved to the Institute of Animal Genetics at Edinburgh University.[61] The institute was something of a hub for unconventional approaches to genetics and was no stranger to graft hybridization. Back in 1925 its director, the geneticist Francis Albert Eley Crew, had experimented with grafting ovaries into different types of guinea pigs. Based on unexpected changes to the color of their fur, he had concluded that graft hybridization could produce small, hereditary changes.[62] By the time McLaren entered the institute it was run by the unorthodox embryologist Conrad Hal Waddington. He was interested in grafting as an experimental technique but gave no indication that he believed in graft hybrids.[63] Waddington also leaned toward the political left, but sometimes became frustrated with overtly political colleagues. When he was approached by the Marxist biochemist Joseph Needham with an offer to join the Society for Anglo-Chinese Understanding, Waddington did so on "the basis that this will not be a narrowly communist society, as far as this country is concerned."[64]

McLaren found a "wonderful scientific atmosphere" at the institute, and remained in Edinburgh for the next fifteen years. She later returned to University College London before moving to the Wellcome Cancer Research Centre (later the Gurdon Institute) in Cambridge.[65] In 1985 McLaren briefly visited Beijing to deliver a series of lectures and visit leading biological laboratories. She noted that the Developmental Biology Institute had yet to settle on a research direction, as "one of the Director's problems is that two of the major lines of work are generally not accepted in the West." She was referring to work conducted at the institute by the late T. C. Tung, who had claimed to have found "inherited cytoplasmic influences following nuclear transfer between fish eggs."[66] New developments in biology and biotechnology only reinforced McLaren's belief in the inheritance of acquired characters. In 1992 she attended a discussion at the Royal Society on transgenic petunias.

Shortly afterward, she wrote another defense of graft hybridization, but also remarked that genetically engineered flowers "made poor Glavinic's tomato results seem positively boring."[67] A famous 1999 article, "Too Late for the Midwife Toad"—its title a reference to the Paul Kammerer controversy—drew upon the epigenetic work of Waddington. In it, McLaren argued that molecular biology demonstrated that environmental stresses on certain molecules could lead to a greater diversity of gene expression.[68]

It is tempting to see the late twentieth-century Anne McLaren as a kind of modern Lamarckian or epigeneticist, who promoted the importance of the environment in an organism's development. Yet the legacy of Lysenko and the Cold War never truly left her. At a 2003 lecture, given in Mendel's abbey in Brno, McLaren spoke favorably of Michurin and the Lysenkoist botanist Glushchenko. The latter's "detailed accounts of the sexual transmission of altered characteristics acquired by grafting are convincing," she claimed, "and have subsequently been confirmed by several workers in Japan and China as well as Europe."[69] One point of interest from McLaren's lecture was her reference to scientific work in Japan. This reference was based upon the research of Yasuo Ohta, a member of the Institute of Agriculture and Forestry at the University of Tsukuba. Ohta had first speculated that genes could be exchanged between grafted plants in 1970.[70] A few years later, in collaboration with a Vietnamese colleague, he sought to test this hypothesis. Taking two varieties of pepper, one red and one yellow, Ohta grafted them together and pruned the leaves of the scion, leaving it dependent on the stock (the Michurin "mentor" method). This, Ohta reasoned, would ensure a "one-way flow of genetic material from stock to scion."[71]

Ohta raised self-fertilized seeds from the stock and scion of his grafted peppers. Some of these young plants gave rise to fruits consisting of characters from the red and yellow pepper varieties. One of the more remarkable examples gave red "pendent" fruits, which hung from its branches. The red pepper variety used by Ohta and his colleague had only ever given upright fruit. Pendent fruit was a characteristic observed only in the yellow pepper variety. "We never had such plants in the experimental field," noted Ohta, "with fruits red yet pendent, non-fasciculate and K-shaped."[72] He argued that these changes could not be the result of contamination or spontaneous mutation, as control groups of ordinarily bred peppers had not yielded this unique set of characteristics. Unlike many of the graft hybridizers we have encountered thus far, Ohta was able to continue his experiments for multiple generations. When his grafted plants were crossed with ordinary peppers, unusual variations could still be observed in their offspring. "The experiments clearly revealed," concluded Ohta, that "grafting in red peppers has induced hereditary changes of certain Mendelian traits."[73] In a 1977 paper he would go on to track the movement of chromatin (the material that makes up chromosomes) across the junction of grafted plants.[74] It was a reissued version of this paper that caught McLaren's eye.[75]

If McLaren had focused her energy on describing these experiments and ongoing work on graft hybrids in Japan, her 2003 Brno lecture may well have been more palatable for Western scientists. But she still had an ideological axe to grind. McLaren used much of her lecture to describe her impression of Lysenko after the 1957 interview with Michie in Moscow. She recalled how she had traveled to the Moscow Institute of Genetics with Michie and encountered Lysenko. Her impression of the infamous agronomist was highly positive. "Lysenko's political activities, and their serious consequences for the future of Russian genetics, have largely obscured the positive aspects of his scientific contribution," explained McLaren. "Lysenko himself was intelligent and thoughtful, with interesting ideas."[76] She argued that epigenetic inheritance implied that the inheritance of acquired characters could also occur. McLaren also suggested that modern Mendelians, such as C. D. Darlington and Thomas Hunt Morgan, were far stricter than Mendel by dismissing the idea that "hereditary factors might exert a reciprocal effect on one another." If Mendel would not have agreed with modern Mendelians, McLaren apparently saw no reason why she could not place him on her side of the Cold War divide in biology. This logic led her to conjure a truly surreal scene:

> Lysenko often criticized (in characteristically Russian forthright terms) the views of T. H. Morgan, and he often spoke disparagingly of 'Mendel-Morganism,' but to the best of my knowledge he never criticized the work or views of Mendel himself.
>
> Of course Mendel never met either Michurin or Lysenko. . . . If through some time-warp they had visited him in his monastery, he would have shown them round his beloved garden, introduced them to 'his children,' and they would have embarked on a most amicable discussion of the work and views of Charles Darwin.[77]

There is no doubt that—given access to a time machine—such a friendly scene might have occurred. If, however, either the Mendel or the Michurin in this fictional meeting had been aware of future events, they might have been more disgusted than pleased by Lysenko's arrival. McLaren was not successful in finding a place in which to publish her Brno lecture. "When I hear terms like 'Mendel-Morganist,' but equally 'Lysenkoist,'" wrote McLaren in response to one setback, "I suspect that politics is getting in the way of scientific inquiry." Graft hybridization, in her view, was now supported by "a sufficient number of positive results, mainly in Japan, including some molecular data."[78] In this instance, the editor at *Trends in Genetics*, Robert Shields, was not convinced. The global reception of Lysenko in the late twentieth century was clearly divided along national and ideological lines. Shields told McLaren, "I am fairly bombarded by suggestions from China that I publish letters saying Lysenko was not all wrong."[79]

One of the most vocal advocates of Lysenko in China was Yongsheng Liu, a plant breeder based at the Henan Institute of Science and Technology. When Liu encountered an article on Hašek's transplantation experiments, he recognized the influence of "Michurin's principles of plant grafting" and was put in touch with McLaren in 2003. Liu professed to be interested in the heredity of fruit trees and the possibilities of "tree species conversion" by grafting. McLaren, by this time well accustomed to international exchanges on graft hybridization, responded encouragingly. "The whole topic of graft hybridization," she wrote, "the inheritance of acquired characteristics and epigenetics as well as genetic changes, is one of very great interest at the present time." In the course of their exchange, Liu would be further inspired by his reading of McLaren's "Too Late for the Midwife Toad."[80]

The result of Liu and McLaren's collaboration was a series of articles by Liu arguing in favor of graft hybridization during the early years of the twenty-first century. A 2006 article in the journal *Advances in Genetics*, for example, "reconsiders the subject of graft hybridization in light of our present understanding." Liu declared that graft hybridization "is compatible with concepts of molecular genetics," and that "graft hybridization and sexual hybridization can coexist comfortably in the universe of Darwin's Pangenesis and molecular biology."[81] In true Marxist fashion, establishing a respectable history for graft hybridization was almost as important for its modern supporters as anchoring its existence in science. Liu's paper invoked many of the historical actors we are now familiar with—including Darwin, Winkler, and Michurin—to support the validity of graft hybridization.[82] The paper also drew upon evidence from Lysenkoists from the 1940s to the 1970s, including French Michurinists.[83] Based on this and other papers, Liu essentially arrived at two key conclusions:

1) Darwin's pangenesis is our most compelling theory of heredity to date. In 1865 Gregor Mendel presented his now-famous paper on experiments in plant hybridization to the Natural History Society of Brno. Ignored in its own time, Mendel's paper was rediscovered at the turn of the twentieth century, with Mendelian genetics subsequently emerging as a cornerstone of modern biology, or so the textbook story goes. Yet Liu argued that Mendelian laws of inheritance are inadequate, as they cannot explain how heritable material can cross graft junctions to form graft hybrids—a phenomenon that also allows for the inheritance of acquired characters. Our best explanation for graft hybridization, Liu argued, is found by looking beyond Mendel and (re)embracing Darwin's theory of pangenesis. Liu's favored mechanism for the formation of graft hybrids was messenger RNA: molecules made up of nucleotides, which are supposedly capable of traveling between grafts.[84] Liu argued that messenger RNA and Darwin's hereditary particles—gemmules—were analogous. In a 2004 paper Liu argued that Darwin's gemmules now have a biochemical basis. In Liu's own words: "Once most geneticists have recognized the existence of graft hybrids and Darwin's so-called gemmules, Pangenesis needs to be reconsidered."[85]

2) We would be better off with a Michurinist take on plant breeding. As we have seen, the Russian horticulturalist Michurin was one of the key proponents of graft hybridization during the early decades of the twentieth century. Faced with the longstanding problem of why grafted fruit trees produced an inferior crop, Michurin theorized that unreliable fruiting occurred when cultivated trees were grafted onto wild stock: unwanted characteristics were, unknown to horticulturalists, being exchanged across graft junctions. Michurin went on to develop the "mentor-grafting" method, attaching cuttings from young seedlings to mature fruit trees, which he claimed to have used to develop several graft hybrids. Michurin's work possessed many attractive qualities for Liu: Michurin was a firm believer in the graft hybrid and announced that Mendelian laws—while useful in many aspects of breeding—could not be applied to fruit trees.[86] Liu was also skeptical toward most modern forms of genetic biotechnology, including recombinant DNA technology, protoplast fusion, tissue culture, and mutation breeding. He declared that graft hybridization offered the best means of changing "quantitative characters" in crop plants. It would reduce the amount of time needed to produce new varieties and allow breeders to transfer select genes from "relatively distantly related species."[87] In other words, graft hybridization is "a simple and powerful means of plant breeding," of great practical benefit, and can accomplish anything that other forms of plant biotechnology can.[88]

After the deaths of Donald Michie and Anne McLaren in 2007, Liu revealed more of the role of McLaren in developing his defense of graft hybridization. Liu, who had been taught both Mendelian and Michurinist genetics, had been attempting to write an article reviewing the scientific literature on graft hybridization. However, his early drafts had been rejected by several journals. Liu recalled that McLaren was sympathetic: "She knew that there is convincing published literature on heritable changes induced by grafting. She suggested that I write a longer review article, with more of the published evidence."[89] Her input likely accounted for some of the character of Liu's 2006 paper in *Advances in Genetics*, with its references to historical figures such as Darwin and Winkler. Liu continued to publish on graft hybridization after McLaren's death, with a 2022 article discussing "interfamily grafting" in plants.[90] Given Liu's interest in rehabilitating Lysenko, it is safe to say that an ideological component exists behind some contemporary research on graft hybrids.

The Graft Hybrid Goes Global

"History may be circular," wrote McLaren in 1999, "but the history of science is helical: it repeats itself, but each time at a deeper level." She used this phrase to convey the idea that the debate over the inheritance of acquired characteristics had reopened in the modern era. This time, however, the debate would be conducted in the language of molecular biology.[91] By this point McLaren had been fighting for the recognition of the inheritance of acquired characteristics for almost half a century. At the heart of her battle against the limitations of genetics was the graft hybrid. There were sound scientific and political reasons for McLaren to focus on graft hybridization. From a political perspective, the graft hybrid had become an integral part of Lysenko's biology. Its existence had become bound up with communism and Soviet biology. When McLaren and Michie met with a weakened Lysenko in 1957, however, this bond had loosened. Lysenko had abandoned the idea of crossing species through grafting and downplayed the importance of the technique for his biology.

Scientifically speaking, however, the graft hybrid was the most palatable part of Lysenko's biology. The possibilities of the technique had given rise to a long-running debate in the West, with the graft hybrid question never being satisfactorily resolved. The well-traveled and connected McLaren was able to gather fresh evidence in favor of graft hybridization from around the world. She had been instrumental in bringing the intriguing and unexplainable experiments of Milan Hašek in Czechoslovakia to the attention of her Western colleagues in the early 1950s. Shortly after her meeting with Lysenko, she traveled to Yugoslavia in a failed attempt to repeat tomato-grafting experiments there. She also attempted to acquire more information on graft hybrids in China. By the late twentieth century McLaren was relying heavily upon grafting research in Japan to support her case. Her collaboration with Yongsheng Liu in the early years of the twenty-first century was the latest iteration of a pattern of internationalist science dating back some fifty years. While the graft hybrid had political baggage, it was undeniable that it now had a global reach and a global history. Over the course of the twentieth century, graft hybridizers had appeared in France, Britain, Germany, Hungary, Czechoslovakia, Yugoslavia, Russia, China, Japan, and the United States.

An array of international sources and contacts was available to McLaren, which allowed her to continually push against the Western limitations to heredity for decades. Why did her efforts not make more of an impact? Part of the reason was the flexibility of genetics. Surprising experimental results, which could be attributed to graft hybridization, were often subsumed into Western biology through concepts such as "acquired immunological tolerance" or the movement of viruses across plant grafts. McLaren would also find that repeating or verifying graft hybridization was not easy. Her collaboration with Ružicka Glavinic, for instance, did not produce the desired results. She also became interested in graft hybrids at the very moment that many researchers were abandoning the subject. This was the case in China, where Teh-Ming Tsu halted his grafting experiments on eggplants as Lysenko fell out of favor. Still, as the decades progressed, McLaren was able to find new sources supporting graft hybridization. From Japan, Yasuo Ohta produced compelling experiments. From China, Yongsheng Liu injected new energy into the Lysenkoist argument for graft hybrids.

All this, however, took place in a rapidly changing world. The graft hybrid had now entered the world of modern biotechnology, which also promised to bridge the gap between species to produce new plants and animals. On a practical and economic level, graft hybridization now had some serious competition.

6

Cell Fusion

—ℓ

In November 1970, the radical politics of the 1960s combined with cutting-edge advances in biotechnology. The venue of this convergence was a meeting of the British Society for Social Responsibility in Science (BSSRS) in London. The society was essentially the public face of a "new scientist–activist movement," which feared that the abuse of science—particularly in pursuit of new weapons of war—would destroy the future.[1] Yet some participants in this 1970 meeting, titled "The Social Impact of Modern Biology," struck an optimistic tone. Addressing a mixed audience of scientists, historians, technicians, and social radicals, Yale's professor of biology Arthur W. Galston declared that it "has now been convincingly demonstrated that animal cells of widely divergent genomes can be caused to fuse somatically."[2] Recent scientific advances had also seen the same feat accomplished in plants. "One can dream of many exciting possibilities," claimed Galston, when considering a future dominated by "new somatic genetics of higher plants." If desirable characteristics could be freely exchanged between different plant varieties or species using cell fusion, this future might include crops that could acquire their own nitrogen from the atmosphere, bypassing any need for artificial fertilizers, or plants that could fight off any disease or fungi that threatened them.[3]

With all this talk of fusing different cells to create a world of limitless possibilities, Galston sounded very similar to the most ardent of the graft hybridizers. He was not referring to the possibilities of grafting, however. He was instead speaking of a technique called "somatic hybridization," or the fusion of plant cell nuclei.[4] Using this technique, biologists were able to strip plant cells of their tough cell wall to create "naked cells," or "protoplasts." These protoplasts could then be combined to create interspecific crosses containing a huge range of genetic information. For almost a decade, it seemed that somatic hybridization would become the leading technique in plant biotechnology. This was the height of "Cold War technological optimism," a time when plant physiologists had claimed not only to have achieved their goal of unlocking the underlying laws of plant physiology but to have overcome the barrier posed by the plant cell wall.[5] The removal of this wall promised the ability to study plant cells with newfound clarity, to merge these cells through somatic hybridization and bypass the limits of traditional sexual reproduction. As late as the 1970s, a future dominated by genetic modification through recombinant DNA technology was by no means a foregone conclusion.

Although somatic hybridization was cloaked in the language of molecular biology, the technique and the promises associated with it were very similar to graft hybridization. Both involved merging the somatic cells of the body, rather than the sex cells. By this means, practitioners of both somatic and graft hybridization hoped to overcome the natural barriers to sexual hybridization. There were structural and institutional similarities too. While the development of recombinant DNA technology took place in a highly commercialized and localized context, research into somatic hybridization occurred in an international academic setting.[6] It was dominated by plant physiologists and pathologists, who considered their disciplines best able to "study and explain biological functions and processes."[7] Overall, somatic hybridizers were rather similar to the graft hybridizers of Western Europe during the 1920s and 1930s. These early twentieth-century botanists and plant physiologists also worked in a largely academic setting, at colleges and universities. They also pursued basic research on how plants functioned and responded to their environment, all the while chasing promising leads on graft hybridization. One key variance that separated the two groups, however, was the Lysenko affair and its polarizing impact in the biological sciences.

There are two good reasons for uncovering the history of cell fusion and its practical applications here. Firstly, it is undeniable that cell fusion and graft hybridization have fundamental similarities. The former, in fact, could be characterized as grafting on the cellular level. Secondly, these similarities were recognized across the twentieth century. Early attempts at cell fusion were not far removed from grafting. In fact, some of the graft hybridizers we have already encountered also attempted to perform cell fusion, or speculated that it was the mechanism behind the emergence of graft hybrids. During the 1960s and 1970s somatic hybridization broke through a series of technical milestones and even attracted the attention of geneticists. Yet its similarities to graft hybridization drove some biologists to actively distance their work on protoplasts from grafting and Lysenkoism. Ultimately, somatic hybridization did not achieve the same level of uptake as other plant-breeding technologies. Although cell fusion had first offered the means to revolutionize agriculture, it was recombinant DNA technology that was able to quickly produce commercial crop plants.

Exposing the Cell

A key moment in the modern history of somatic hybridization occurred at the University of Nottingham's Botany Department in 1960. Some forty years later Edward C. Cocking, its principal instigator and a lecturer in plant physiology, recounted the event. Cocking was attempting to develop a new cell culture method. Noting that cell division did not occur in tomato root cells, he speculated that releasing the cell contents from their confining wall would aid the culture process. Drawing upon discussions with workers at the Microbiological Research Establishment in Porton, England, Cocking decided that a cellulase enzyme would be most effective for degrading plant cell walls. Fruitless attempt after fruitless attempt followed his decision. Commercially available enzyme preparations were simply not up to the task. A promising avenue finally opened when Cocking came across the studies of D. R. Whitaker, of the National Research Laboratories in Ottawa, Canada, who had developed his own cellulase preparation. When Cocking tested Whitaker's preparation, the solution was a complete success, releasing protoplasts.[8]

To understand the importance of Cocking's chemical dismantling of the cell wall, we must step back into the history of efforts to remove the wall of plant cells and fuse their contents. Although the first recorded protoplasts were observed in the late nineteenth century, early milestones in what would later be termed *somatic hybridization* were disconnected from more modern developments. These milestones were recognized as significant only following

FIGURE 12. Nineteenth-century equipment for studying protoplasts. In 1892, John Klercker, a botanist at the University of Stockholm, managed to remove the wall of plant cells to produce what would later be termed protoplasts. A chemical method of removing the cell wall was devised by plant physiologist Edward C. Cocking in 1960. Klercker, "Eine Methode zur Isolierung lebender Protoplasten," *Pflanzenphysiologische Mitteilungen* 3, no. 9 (1892): 467.

reviews of the scientific literature from somatic hybridization enthusiasts during the 1960s and 1970s. It was such reviews that uncovered the work of John Klercker, an associate professor of botany at the University of Stockholm, who in 1892 had mechanically cut away the wall of plant cells to release their cytoplasm and observe their contents (fig. 12).[9] Cell fusion between protoplasts was subsequently observed in epidermis cells by the German botanist Ernst Küster in 1910.[10] In the midst of the First World War, Hans Winkler—graft hybridizer and creator of the famed *Solanum darwinianum*—also began working on plant cells. Using the dehydrated tissue of two *Solanum* species, Winkler was able to create tetraploid cells (with four times the usual number of chromosomes). His cell fusion was intimately tied to his interest in grafting. He attempted to use grafting in tandem with his cell work to produce plants through fusion. One of his results was a giant nightshade plant with 144 chromosomes.[11]

The desire to peer inside the cell would engage plant physiologists from across the world. But mechanical methods of removing the cell wall were extremely difficult and labor intensive, severely limiting the number of protoplasts available for study. Writing in 1931, Janet Q. Plowe, of the University of Pennsylvania's Department of Botany, described the agonizing process of separating dehydrated epidermal cells of Bermuda onions from their walls using nothing more than a blunt needle and a scalpel.[12] It is worth reiterating at this stage that the early pioneers of protoplast creation and fusion were not capable of creating somatic hybrids. They were plant physiologists engaged in basic research within university botany departments. As demonstrated by Plowe's paper, which explained how "the existence and function of the plasma membrane" concerned physiologists "from both a practical and . . . theoretical point of view," their interests were focused squarely upon the plant cell: its structure and function.[13] It would not be until the 1960s that the means of producing large numbers of plant protoplasts—hence raising the possibility that somatic hybrids could be a useful tool in plant breeding—would become available.[14]

The interwar period did see indications that some kind of cell fusion might one day play a role in plant breeding. In the late 1920s Küster observed naked vacuolar membranes in the sap of solanaceous berries. This suggested that protoplasts could occur naturally in some plants, raising the tantalizing possibility that cell fusion could follow.[15] Cell fusion between different plant species was achieved in 1937 by Küster's protégé, W. Michel, who used a sodium nitrate solution to encourage protoplasts to merge. However, Michel faced many of the same problems encountered by graft hybridizers. Fusion events were very rare and he was unable to culture more cells, let alone grow new plants, from his fused protoplasts.[16] Nor did his achievement influence the graft hybrid debate of his era. In 1934 William Neilson Jones had felt unable to declare that graft hybrids existed, as "hitherto no certain evidence has been offered that vegetative [or somatic] fusions between genetically dissimilar neighbouring cells do in fact occur."[17] There is no indication that Neilson Jones was ever alerted to Michel's demonstration of plant cell fusion.

The priorities and objectives of various scientific disciplines aligned with cell fusion throughout the twentieth century. Plant physiologists, for instance, insisted on the primacy of their discipline within botany. By the 1950s this "self-image" manifested itself with a focus on the basic processes underpinning life, an experimental methodology, and a belief that plant physiology was the "leading edge of plant science."[18] Despite this sense of primacy and an experimental drive, by the 1950s protoplasts existed only as a rare research tool for plant physiology. Cell fusion was of great interest to biologists engaged with fundamental questions surrounding heredity and the plasticity of life. During the mid-twentieth century, new discoveries indicated that somatic cells could exchange genetic information, leading molecular biologist and bacteriologist Joshua Lederberg to criticize biologists for their "antisexual bias."[19] Cell fusion was also a useful tool for studying the genetics of bacteria and fungi. The hybrid cells created by fusion allowed researchers to study how the genes of parent microbes were expressed and regulated.[20]

Creating hybrids from the somatic cells of so-called higher organisms was more difficult. The plant cell had its tough cell wall. Animal cells were also resistant to fusion, a barrier overcome in 1953 when biomedical scientist John Franklin Enders found that viruses could cause cultured cells to fuse. In 1962, a group of biologists headed by Yoshio Okada, a virologist at Osaka University, exposed a virus to radiation and used it to fuse mouse cells.[21] Okada's work on cell fusion won him the Annual Prize of the Japan Society of Human Genetics in 1978. His acceptance speech blurred the lines between the somatic cells of the body and the sex cells. "In classic genetics," explained Okada, "an individual animal body is understood as only the phenotype of the genomes of a fertilized egg. According to this theory, a somatic cell is only one member of the whole body." The ability of biologists to grow somatic cells in the laboratory, however, meant that it was possible to "handle a somatic cell as an autonomous individual in a Petri dish." Since genetic information from both parents was contained within the cells of animals, Okada concluded "that a somatic cell at the level of genomes is equivalent to a fertilized egg. Thus, we can use a somatic cell *in vitro* for genetic analysis instead of a whole body."[22] His speech may have challenged the Weismann barrier, but it did not break it. After all, at the time there was no way for scientists to grow an animal from fused cells and see if its hybrid genes would be inherited by its offspring.

Henry Harris and John Watkins, two Oxford-based biologists, stole the cell fusion limelight when they merged human and mouse cells in 1965.[23] Their decision to fuse our species with another was deliberate, calculated to "draw attention to their work."[24] Harris situated their experiments firmly within the medical tradition. In a 1966 lecture to the Royal Society, he described how the relationship between certain diseases and fused cells was frequently discussed in the medical literature of the nineteenth century.[25] Its shock factor aside, the fusion of human cells was useful for understanding the workings of cell genetics, including those responsible for certain cancers. Harris did eventually make a brief reference to plant grafting in the early 2000s, when he translated an early twentieth-century treatise on cancer by the German zoologist Theodor Boveri. Faced with an ambiguous reference to grafting, Harris speculated, "Perhaps he is referring to the vegetative propagation of plants, but, if so, it is not easy to see a relevance to the cancer problem."[26] This passage would seem to indicate that the pioneers of animal cell fusion were either unaware of the graft hybrid controversy or chose to ignore any similarities with their own research.

The successful fusion of human and animal somatic cells left biologists surprised by the cellular compatibility, or "internal homology," of organisms.[27] Fusion of animal cells now served as a source of dialogue and inspiration for those involved in somatic hybridization in plants and microorganisms. Everyone who practiced cell fusion was wholly reliant upon tissue culture for their work, with techniques readily shared across disciplinary boundaries throughout the twentieth century.[28] As somatic hybridization developed throughout the 1960s and 1970s, new innovations were passed on to colleagues concerned with animal cell fusion. For instance, Harris recalled how a highly effective chemical that was used to encourage plant cell fusion in the mid-1970s was also found to be of equal benefit for fusing animal cells.[29]

To journalists, the fusion of human and mouse cells heralded everything from the creation of monsters to a new understanding of life.[30] Cell fusion faced a mixed public reception. Watkins criticized a 1967 BBC television program titled "Assault on Life" for not revealing the medical purpose behind his experiments with hybrid cells. Responding to a distressed viewer of the program, Watkins explained that cell fusion was not the same as fusing animals, stating that "We are not, for example, trying to create centaurs." Nor would cell fusion create "species of subnormal intelligence," or "ferocious species in invincible armies." In a manner very different from that of his publicity-courting colleague Harris, Watkins stated that the public reaction to the BBC program provided an "awful warning" to scientists tempted by the world of "mass media, fashion photographers and pop stars."[31] For better or worse, cell fusion had become part of the "Biological Revolution." In the February 1969 issue of the *Atlantic Monthly* magazine, Harvard historian Donald Fleming announced that the biological revolution was an event "likely to be as decisive for the history of the next 150 years as the Industrial Revolution has been for the period since 1750." Among the achievements of this revolution had been the discovery of the structure of DNA and the successful use of organ transplants in medicine. Now Fleming could add the creation of hybrid cells via the fusion of different species.[32] Scientific reaction to cell fusion was just as enthusiastic, with one anonymous contributor to *Nature* describing it as "a new gift to biology" in 1969. Looking for an analogy with which to describe the new technology, they settled on not grafting but fertilization. "Cell fusion," they explained, "is like fertilization, except that it is an unnatural kind of union that has no right to occur." Although fusion could breach the species barrier, its main utility was to the study of heredity: "Furthermore, certain hybrid cells are a godsend to geneticists because of their propensity for shedding chromosomes. At fusion, the hybrid cell possesses the genetic endowment of both its parent cells, but in subsequent cell divisions various chromosomes are discarded and with them the properties they conferred on the cell. Short of being able to breed mammals as thick and fast as fruit flies, the hybrid cell offers an unrivalled opportunity for genetic analysis."[33] From this perspective, cell fusion in the laboratory was like a miniaturized breeding experiment. Attempting to breed animals on a large scale to analyze their genetics was expensive and time consuming. Cellular work could accomplish in weeks what would otherwise take years. Intriguingly, cell fusion was now held up as a "godsend" to geneticists, rather than regarded as a controversial term connected to graft hybridization. Of even greater interest was the finding that the malignancy of cancerous cells could be suppressed following fusion, raising hopes of a revolution in medicine.[34] "During the past five years," concluded the anonymous contributor, "cell fusion has grown from being something more than a curiosity to a technique of considerable power and flexibility." Despite the promise of the technique, however, "the results of cell fusion require considerable skill, sweat and luck to interpret."[35]

The excitement that cell fusion generated in the scientific community led to international collaboration, even across the Iron Curtain. George Klein, a Hungarian microbiologist, experienced this firsthand. As a young medical student, Klein had managed to emigrate to Sweden and attended the 1948 International Congress of Genetics in Stockholm. Here, he heard the American geneticist Hermann Joseph Muller denounce Trofim Lysenko as a charlatan and offer his resignation to the Soviet Academy of Sciences. Klein also heard about advances in microbial genetics. "Bacteriology had been the last citadel of Lamarckism," he later recalled.[36] Now his research on cancerous cells shifted to population dynamics, mutation, and evolution.

Almost two decades later, Harris reached out to Klein's cancer research team in Sweden with the suggestion that they conduct experiments using hybrid cells. The team was also joined by Francis Wiener, a Hungarian-Romanian pathologist who worked with Harris to fuse cells with different levels of malignancy. They concluded that tumorigenicity—the tendency of cells to form tumors—"was suppressed by fusion with normal cells. It reappears after some critically important chromosomes, contributed by the normal cell, have been lost."[37]

Amid the revolutionary chaos of animal cell fusion, somatic hybridization of plants was gradually gaining pace as an exciting new technique. Cocking's 1960 report to *Nature* on his chemical method of releasing protoplasts was purely descriptive. Nevertheless, a paper published the following year showed developments in both his techniques and fresh ideas on the potential use of protoplasts. Cocking noted that "liberated bacterial and fungal protoplasts" were already of great value in "morphological, biochemical and genetic work." Protoplasts released from the root tips of tomato seedlings in Cocking's laboratory "indicated their unique potentiality for similar studies."[38] An enzymatic means of creating protoplasts also freed physiologists from the constraints of micromanipulation of cells via surgical instruments as described by Plowe in 1931. Relatively speaking, protoplasts could now be created quickly and in the large numbers required for research. In these early years, however, experiments conducted at the University of Nottingham focused purely upon means of harnessing protoplasts to solve "present problems associated with growth and differentiation in plants."[39]

The twentieth century had seen a growing confidence among plant physiologists that their experimentally oriented discipline could unlock the fundamental processes of life. Part of this ambition manifested itself in attempts to remove the plant cell wall and study protoplasts, as attempted by Klercker and Plowe. Yet to some extent this history of protoplast creation was an invented tradition. In 1967 controversy erupted when Swiss botanist Albert Frey-Wyssling suggested that plant protoplasts should be termed *gymnoplasts*. Frey-Wyssling based his challenge upon historical precedence, citing Küster's use of the term.[40] Unfortunately for Frey-Wyssling, his claim to historical precedence using Küster was subsequently overridden by the (re)discovery of Klercker's earlier 1892 manuscript. Disputes over terminology can be seen as part of a more significant struggle to construct a scientific tradition. As protoplast research dramatically surged forward during the 1960s, the creation of commercially important somatic hybrids became a tangible possibility. The recognition of who came first suddenly became a matter of urgency.

One of the contributors to the 1965 *Encyclopedia of Plant Physiology* was F. Brabec, who announced that "somatic hybrids do not exist and taking all possibilities into consideration, it appears unlikely that they will ever exist."[41] The timing of Brabec's statement seems off, given that cell fusion had united human and animal while plant protoplasts had become available in formerly unimaginable numbers. His claim makes sense, however, if we interpret it as a rebuttal of graft hybridization. By the mid-twentieth century, graft hybridization was an unpopular idea among Western biologists. It had long been the source of fantastical claims and was indelibly linked with Lysenko. Now researchers were astonished to find that very different organisms were compatible on the cellular level. By the 1960s the stage was set for the revival and future development of protoplast research with a new aim: the creation and reproduction of somatic hybrids. One of the challenges of this new era would be to differentiate a promising form of biotechnology from Lysenko's grafting.

The Grafting Taboo

Scientists at the University of Nottingham may have been reticent to make wild claims regarding the potential of protoplasts for plant breeding. Other biologists were not so reserved, excitedly noting the potential for somatic hybridization. Speaking at the 1970 BSSRS meeting, Galston embodied this excitement. But why did naked cells in a Nottingham laboratory so excite this professor of biology? In the spirit of a conference skeptical of scientific triumphalism, Galston characterized intensive agriculture as beset by technological problems, from overreliance on fertilizers to disease-vulnerable monocultures.[42] Radical advances in plant breeding would be required to produce new crops: plants capable of yielding more food at a lower cost to the environment. This issue was made all the more pressing by the publication in 1962 of Rachel Carson's *Silent Spring*, which revealed the extent of environmental damage caused by indiscriminate pesticide use in industrialized agriculture.

Only two years before the BSSRS conference, in 1968, Paul Ehrlich published his own bestselling work, *The Population Bomb*. The book warned of the precise dangers to modern agriculture cited by Galston, including the environmental degradation caused by nitrogen fertilizers, while predicting global food shortages from overpopulation. Population concerns would go on to feature prominently at the first Earth Day in 1970.[43] On the one hand, industrialized agriculture was a source of pollution and environmental damage. On the other, growing more food was one way to counter the looming population crisis. Among a number of promising solutions discussed by Galston was somatic hybridization. Referring to Cocking's removal of the plant cell wall, Galston announced that somatic hybrids might one day emerge, possessing remarkable qualities from nitrogen fixing to disease resistance.[44]

Within purely scientific exchanges, a similar level of excitement was displayed. At a 1969 meeting of plant physiologists and geneticists, somatic hybridization was designated by observers to be "still experimental," but showing "great promise." Suggestions arose that sexual barriers to crossing in plant breeding could be overcome. Advances in protoplast manipulation hinted that "asexual fusion might become a major method for 'crossing' unrelated plants which are not easily crossed using standard sexual methods."[45] Reported in *Science*, the meeting "Crop Improvement through Plant Cell and Tissue Culture" was no minor affair and included important figures such as Cocking.[46] Despite the sanguinity of the attendees and Galston's optimism, though, it had now been some ten years since Cocking had first harnessed enzymes to release protoplasts. Not one plant had yet been created using somatic hybridization.

Two barriers stood in the way of somatic hybrids. Once released from the confines of their cell wall, protoplasts were no longer viable as living cells outside of their nurturing medium. Vulnerable to the environment, the regeneration of a new cell wall was necessary for their long-term survival. With this achieved, efforts could then turn to growing viable plants from protoplasts. These barriers were overcome due to the efforts of Japanese researchers. In 1970 Toshiyuki Nagata and Itaru Takebe, of the Institute for Plant Virus Research in Chiba, Japan, observed protoplasts regenerating their lost walls. Their subject, tobacco mesophyll, was also capable of cell division.[47] Takebe was hopeful. Citing then unpublished observations, he stated his belief that protoplasts were capable of fusion, offering "a unique experimental material for plant genetics."[48] Collaboration between Takebe and researchers at the Max Planck Institute for Biology in Tübingen, Germany, the following year saw the regeneration of a whole plant from protoplasts. These results established "for the first time that cell protoplasts from the mesophyll can be cultured to give rise to whole plants." Extensive cell division in protoplasts opened up new possibilities, including "the breeding of new plants through somatic hybridization."[49]

Given the preexisting interest of plant physiologists in protoplast work, it comes as little surprise that an enzymatic means of releasing protoplasts was first developed in a botany department. Cocking's work raised much interest, with funding for future research provided by esteemed bodies such as the Royal Society and the British Government's Department of Scientific and Industrial Research.[50] The regeneration of whole plants from protoplasts marked another important step toward a new future in plant breeding, one dominated by somatic genetics. These results also emerged from the plant sciences sector, albeit a plant pathology institute rather than a department of botany. The shift in protoplast research to Japan is explained by the country's advanced enzyme-production facilities.[51] The cellulase enzyme used for protoplast production was produced commercially in Japan from 1968, enabling domestic researchers' easy access to the raw ingredients necessary for advanced work with plant protoplasts. Although Takebe and his colleagues had openly invoked the possibility of somatic hybrids in 1971, the actual regeneration of a higher plant from fused protoplasts would take place the following year in the United States.

In 1972—the year of the production of the first recombinant DNA molecules—a team headed by Peter S. Carlson at the Department of Biology at Brookhaven National Laboratory used protoplast fusion to create a plant hybrid between two varieties in the tobacco genus: *Nicotiana glauca* and *Nicotiana langsdorffii*. This achievement marked a major advance in the field of somatic hybridization, moving the fledgling technology one step closer to its ultimate commercial aim: creating new varieties of enhanced crop plants in agriculture. Human manipulation had overcome the sexual barrier to species crosses. The Brookhaven National Laboratory team's paper, published in the *Proceedings of the National Academy of Sciences*, bore clear references to the difficult and terminologically confusing past of somatic hybridization. Nagata and Takebe's experimental conditions and regeneration medium were also exactly re-created. Although the Brookhaven team had succeeded in encouraging a hybrid cell to develop into something resembling a plant, its growth had not been a complete success. The fledgling hybrid grew shoots and leaves in its tissue culture, but had no roots. Its creators were therefore forced to resort to grafting. "In order to obtain further differentiation of presumed hybrid tissue," they explained, "the regenerated shoots were grafted onto the freshly cut stem surface of young plants of *N. glauca*."[52]

Grafting effectively saved the young somatic hybrid. Its leaves continued to develop and the Brookhaven team were able to take samples for their

1972 publication. But what was the cost of introducing grafting to their experiment? By saving their plant from a premature demise, the team had raised the possibility that grafting might have somehow contributed to its hybrid nature. To combat this impression, a disclaimer subsequently appeared in their account of the experiment: "The somatically produced hybrid spontaneously forms tumorous outgrowths on the stem. . . . Spontaneous tumor formation is a genetically determined trait that is characteristic of the F_1 [first generation] hybrid and amphiploid [an organism with one set of chromosomes from each parent], but is not found in either parent species, and is not transmitted across a graft union."[53] After grafting their sickly hybrid plant onto a stock of young *Nicotiana glauca*, the Brookhaven team observed that tumors spontaneously emerged around the site of the graft junction. This was a "genetically determined trait" that did not occur in either parent of the hybrid. So where had it come from? The Brookhaven team noted that these tumors were also observed in sexual crosses of their tobacco plants. Their somatic hybrid was therefore identical to sexual crosses. Most importantly, this trait was "not transmitted across a graft union." There was no possibility that genes had made their way across the graft junction and that graft hybridization had occurred.

Back in Britain, the success of the Brookhaven Laboratory team featured in the *Financial Times*. The editors of the newspaper's technical section, Arthur Bennett and Ted Schoeters, used the somatically hybridized Brookhaven tobacco plant to announce that "birds and bees [are] not wanted." Cell fusion of different crop varieties had superseded traditional hybridization, avoiding the need for a "complex programme of cross-fertilisation of parent plants." Carlson and the Brookhaven team also claimed that their plants could create "seeds which are fertile and breed true, which is not often the case with the hybrids now being produced by plant breeders."[54] To elements of the British press, somatic hybridization possessed two major positives as a plant-breeding technology. Firstly, fusion of distinct plant varieties and even species could overcome traditional sexual barriers to result in new drought- or disease-resistant crops. Secondly, somatic hybrids offered an alternative to traditional hybrid crops. Not only could cell fusion replicate the results of hybridization but the fertile seeds yielded by somatically hybridized plants could potentially put the ability to replicate seeds back in the hands of farmers.

There was still some way to go before somatic hybridization could create fully functioning crops. The Brookhaven team had managed to raise a somatic hybrid and expressed their hope that the "potential" of somatic hybridization might one day "extend the possibilities of combining widely divergent genotypes of plants." Thus far, however, they had crossed only closely related varieties. "Further attempts to produce a somatic interspecific hybrid and hybrids between more distantly related species in our laboratory," reported the team, "have been hampered by a lack of familiarity with the kind of characteristics the tissue will display."[55] Even with numerous protoplasts now available, breaking the species barrier was not easy. In contrast to the fantastical reports that accompanied breakthroughs in animal cell fusion, Bennett and Schoeters's newspaper article reported that the Brookhaven team "have already discounted science fiction hybrids such as trees bearing immense crops of runner beans all the year round."[56]

By the mid-1970s it had become possible to create somatic hybrids from different plant species.[57] Somatic hybridization was now recognized as an important "breakthrough in cytological and genetical methodology," with some supporters seizing upon its potential to bypass "the limits of traditional plant breeding."[58] Somatic hybridization not only was recognized within scientific circles but also began to inform attitudes to global population and food security. Addressing the Economic Club of Detroit in 1980, Clifton R. Wharton Jr., chancellor of the State University of New York, included somatic hybridization alongside germplasm banks as a future means of combating world hunger.[59] This imagined future was, in fact, some way off. Although plant protoplasts from any species could theoretically be fused, basic cereals and legumes were not easy to raise in tissue culture. Cocking and his team ran into this problem repeatedly. "Indeed, we at Nottingham have spent considerable efforts culturing leaf tissue of wheat, maize and *Sorghum bicolor*," wrote the team in 1977, "but have never obtained a callus"—an unorganized mass of cells, as depicted in fig. 13—"irrespective of the age of the leaf and the composition of the nutrient media."[60] Their experiences convinced them that there was an "urgent need for extensive work in the field of basic tissue culture of cereals and other crop plants before any practical utilization of protoplast technology in agriculture can be achieved."[61]

In 1981 an issue of the *Philosophical Transactions of the Royal Society* titled "The Manipulation of Genetic Systems in Plant Breeding" included a number of articles on somatic hybridization. The issue not only was significant for advocates of somatic hybridization but also discussed numerous breeding techniques and challenges facing contemporary plant scientists and breeders. Cocking noted a marked improvement in the commercial

FIGURE 13. A regenerated cluster of cells, or callus, grown from protoplasts. To make new and commercially viable crops, somatic hybridizers had to encourage their fused cells to grow into plants. The growth of a callus was the first step in creating a young plant. Tina Lorraine Barsby, "Towards Somatic Hybridisation in the Genus Solanum" (PhD diss., University of Nottingham, 1981).

prospects of somatic hybridization, with several horticultural and crop species now having been created through protoplast fusion.[62] He also acknowledged, however, that further research and close collaboration with plant breeders would need to occur before the use of protoplasts (whether through cloning at the cellular level or somatic hybridization) "adds significantly to the armoury of the plant breeder."[63] Geneticist Sir Kenneth Mather was more upbeat, asserting that the main obstacle to the development of new crop varieties through somatic hybridization was the regeneration of whole plants from protoplasts. Recent advances in regeneration and tissue culture made this obstacle less daunting, leading Mather to claim that regeneration from protoplasts would "soon be achieved in our cereals."[64]

As the 1980s progressed, somatic hybridization continued to appear in scientific publications on plant breeding and biotechnology, albeit accompanied by a promising newcomer: genetic manipulation through recombinant DNA technology. The latter became a viable agricultural technology in 1983, with the first permanent uptake of genetic information by a plant.[65] Simultaneous achievements occurred in the production of somatic hybrids, including an infertile cross between a potato and a tomato.[66] In 1985 M. W. Fowler, of the Wolfson Institute of Biotechnology at the University of Sheffield, listed protoplast fusion and genetic manipulation side by side in a review of methods in cell and tissue culture.[67] Even so, for Fowler, somatic hybridization remained a potential tool in agriculture, rather than a practical reality.[68]

In 1984 an international symposium on genetic manipulation in crops was held in Beijing. Xianghui Li, of the Academia Sinica's Institute of Genetics, used his platform to note that somatic hybridization had been hampered by resulting hybrid plants failing to display even a "minimal level of fertility."[69] Technical difficulties dogged somatic hybridization, at the very moment that recombinant DNA technology began to display agricultural applications. Yet all was not lost. At the same symposium, a team comprising members of Agriculture Canada and Carleton University's Biology Department announced practical advances in cultivar creation through somatic hybridization. Working on a tobacco-breeding program, researchers had somatically crossed two varieties, selected for their disease resistance and elevated nicotine levels. Unlike their predecessors, these hybrids displayed usable levels of fertility. Some twenty somatic hybrid lines were transferred from Ottawa to Delhi to be incorporated into a backcrossing program.[70] This line of research finally paid dividends. Some ten years after the symposium, a commercial crop of tobacco created through somatic hybridization was planted in Ontario.[71]

Closing the 1984 symposium, W. R. Scowcroft, of the Commonwealth Scientific and Industrial Research Organisation's Division of Plant Industry, gave his reflections. Scowcroft emphasized the importance of plant biotechnology, which, under his definition, included techniques in tissue culture and genetic engineering.[72] He described the ability to produce large numbers of protoplasts and to induce their regeneration into plants as a "truly remarkable technological achievement." Protoplast fusion was a different matter. Although somatic hybridization allowed "the circumvention of barriers to sexual hybridization," fertility problems meant that it was "still uncertain whether somatic hybridization will permit useful nuclear gene introgression for crop improvement."[73] As genetically modified crops achieved success and provoked controversy on the international scene during the 1990s, news from the world of somatic hybridization was muted. Notable commercial milestones were met during this time, particularly in Canada. The fact remains, however, that somatic hybridization achieved nothing like the status and ubiquity of genetic modification in agriculture.

Despite the technical similarities of graft and somatic hybridization, there was little or no crossover between their supporters and scientific literature during the mid-twentieth century. Part of the reason was likely the taboo surrounding grafting ever since the Lysenko affair. It was almost certainly this Cold War hangover that led the Brookhaven National Laboratory team to explicitly declare in 1972 that genes did not cross between their grafted plants. Another was the nature of somatic hybridization as a field. Like their graft hybrid counterparts, its members were focused on experimental results. They considered the background of their science only in retrospect, situating themselves as the modern descendants of nineteenth-century plant physiologists. Somatic hybridizers embraced this imagined tradition, pursuing the goal of taking somatic hybridization from the laboratory to the field. However, their hopes that cell fusion would become a new tool for plant breeders would run into difficulties, while facing stiff competition from rival recombinant DNA technology.

Theoretically Elegant and Technically Demanding

The story of somatic hybridization is one of unrealized ambition, despite its vast potential. Graft hybrids, too, had once promised to shatter the barrier between species and produce new crops for agriculture. Yet in both cases, practical applications did not readily emerge. Why, then, are fields of somatically hybridized crops absent from our countrysides? The large chronological gaps in the reconstructed story of somatic hybridization offer some indication. If recombinant DNA was a rapidly emerging technology, then protoplast fusion moved at a snail's pace. The technique was later described by British geneticist Norman Simmonds as "theoretically elegant, but technically demanding."[74] But the technical difficulties involved in creating and fusing protoplasts is only part of the explanation. Failure to integrate with existing industrial systems and competition from other methods of plant breeding were significant factors in the slow and halting development of somatic hybridization.

Results from protoplast research came periodically. It was more than a decade after Cocking had first used an enzymatic procedure to create protoplasts that the next step toward somatic hybrids was made: the regeneration of the cell wall of protoplasts.[75] Reflecting upon this gap, Cocking would later describe how his isolation of protoplasts "was ahead of the then technology of plant cell-wall-degrading enzyme production."[76] Shortages of enzyme held back the work of plant scientists at the University of Nottingham. The personal interests of Cocking also held back protoplast work. By his own admission, Cocking was more interested in light microscopy and electron microscopy during the early 1960s, inspired by his work with the biologist Irene Manton at the University of Leeds and the biochemist J. Heinrich Matthaei in Göttingen.[77] Even if large amounts of commercially available enzymes had been available, Cocking considered it "unlikely" that he would have become a pioneer in protoplast fusion.[78]

Cellulase enzyme was made commercially available in Japan in 1968, for the purpose of baby food and biscuit manufacturing. This enabled Japanese protoplast researchers such as Nagata and Takebe to carry out their experiments.[79] Elsewhere, enzyme shortages continued. A 1974 letter from Keith Roberts of the John Innes Institution to James Watson at Cold Spring Harbor Laboratory discussed the possibility of the institution running a course on higher plant cell protoplasts. Despite promising steps in resurrecting somatic plant cells, Roberts identified ongoing difficulties in the field, not least a lack of published literature. The laboratory setup required for a course was relatively simple: a greenhouse, tissue culture facilities, water baths, and bench centrifuges. Roberts noted that cellulase enzymes constituted a significant expense, having to be directly obtained from Japan.[80] As a cutting-edge biotechnology, protoplast production was ahead of existing enzyme-production techniques, and therefore required rare and expensive materials. The development of protoplast research (and hence somatic hybrids) was significantly slowed by enzyme shortages during the 1960s and even into the 1970s.

Technical difficulties with the technology became increasingly evident following the creation of the first somatically hybridized plant in 1972 at Brookhaven National Laboratory. A close reading of the Brookhaven team's paper reveals that somatic hybridization not only was extremely complex but once again ran ahead of existing technology and practices in the biological sciences. Protoplast fusion was not a precise technique. The Brookhaven team found that about a quarter of their protoplasts were actually involved in a "fusion event" (unusually efficient for the time), and even fewer of these contained the genetic information from both parent plants necessary for regeneration.[81] Though an impressive achievement, the somatically hybridized tobacco created at Brookhaven was far from a commercially viable organism. Shoots and leaves developed, but not roots, leading the team to graft their new shoots onto the stems of other plants to further observe the development of their somatic hybrids. Furthermore, spontaneous tumors were observed to develop on the stems of the somatic hybrids.[82] The new plants were delicate and unstable. An equally important and difficult challenge for the researchers was determining whether their new tobacco plants were true somatic hybrids.

Three promising isolates (regenerated plants) were selected for testing to confirm that somatic hybridization had taken place. The Brookhaven team largely relied upon detailed morphological observations, which gave circumstantial evidence that their isolates were somehow different from either parent species.[83] But morphological characteristics could be relied upon only to a certain extent. These characteristics were not necessarily representative of genetic differences and did not indicate exactly which chromosomes had been exchanged between protoplasts. On a practical level, morphology was slow work, requiring researchers to wait for plants to fully develop before the required measurements could be taken.

Other means of determining whether, and to what extent, protoplast fusion had occurred were also used by the Brookhaven team. Electrophoretic analysis demonstrated that the new plants possessed differences in their protein makeup, but electrophoresis did not show which chromosomes had been exchanged, and it was a relatively crude tool for protein fingerprinting of plants by the early 1970s.[84] Extracting chromosomes from the young leaves of the growing plants gave a more definitive answer. These samples contained a chromosome number of forty-two, not unexpected when "the complexity of the fusion event and divisions after fusion" prevented the complete exchange of chromosomes from the parental protoplasts.[85] It was this very unpredictability that led geneticists such as Simmonds to dismiss somatic hybridization as an overly complex biotechnology. Uncertainty and genetic instability caused by the uncontrolled mixing of chromosomes was not an endearing trait of somatic hybridization. Raising somatic hybrids from a cluster of cells to a fully functioning plant represented another difficulty. As we have seen, the Brookhaven team was forced to resort to grafting to save their tobacco hybrid when its roots failed to develop. Cocking and his team at the University of Nottingham also felt let down by the limitations of tissue culture. Common legumes and cereals could not be grown from leaf cells in the laboratory, hindering efforts to apply somatic hybridization to crop plants. The newer recombinant DNA technology did not run into the same barriers.

So far, somatic hybridization has been portrayed as a research topic of international interest, crossing disciplinary boundaries between plant science and genetics with ease. In some cases, however, international collaboration on somatic hybridization was hampered by disciplinary boundaries. A 1984 book by Yury Gleba and Konstantin Sytnik, both at the Ukrainian Academy of Sciences, noted that work on hybridizing somatic cells had been carried out almost entirely by plant physiologists, not plant geneticists. Physiologists had designed methods for cell and protoplast isolation, but an "instillation of genetic ideology and the strict logic of genetic experiments" was needed for further progress.[86] A lack of practical progress and subsequent benefits for plant breeders may have tempered enthusiasm for somatic hybridization. Gleba and Sytnik centered themselves within the biological revolution. Distinct from the "passive" analysis of organisms, somatic hybridization embodied the "synthetic" spirit and purpose of genetic engineering. For Gleba and Sytnik, recombinant DNA technology was in no way superior, as "the results of the [somatic hybridization] experiments . . . force us to believe more and more that the way chosen by their authors for sculpting a novel plant is the efficient one."[87]

The development of recombinant DNA technology is portrayed in some historical narratives as highly focused, in stark contrast to the geographic and disciplinary divides surrounding work on somatic hybridization. The former arose from biotech firms in the United States, the product of a merger of university biology and commerce. Still, commercial links alone cannot completely account for the rise of DNA based technology. Unlike protoplast fusion, recombinant DNA technology was applicable to a wide range of activities in the biological sciences, hence its adoption by "molecular biology laboratories around the world."[88] Somatic hybridization was instead the preserve of plant scientists, leading to the complaints of Gleba and Sytnik. Cocking believed that it was the genetic expertise of the Brookhaven team that allowed them to create the first somatic hybrid; in fact, geneticists initially turned to protoplasts in their quest to modify organisms.[89] Yet a number of factors ultimately encouraged the uptake of recombinant DNA technology as the go-to method for the genetic modification of plants. It was not a simple matter of recombinant DNA being a far easier or more reliable technology, as the creation of these genetically modified plants still involves elements of chance and wastefulness. Recombinant DNA benefited from its place within the rising discipline of molecular biology, leading to widespread interest from both science and industry. This is not to say that somatic hybridization research suffered due to a lack of investment. Cocking, for instance, found himself with sixteen years' worth of funding from the United Kingdom's Agricultural Research Council from 1969.[90]

Somatic hybridization was not the only promising technique to emerge from plant physiologists' investigations into cell fusion. By the 1980s it had become apparent that more often than not, a fused cell would shed one set of its parents' chromosomes (chromosome segregation). The once "considerable hope" that fusion of plant nuclei could result in improved crops faded, to be replaced by interest in "the introduction of small genetic elements from alien species into ones of practical interest"—that is, crop plants.[91] By focusing on the introduction of desirable characteristics from "extranuclear" genes contained within the cytoplasm of cells, cytoplasmic hybrids, or "cybrids," could be created. Back at the University of Nottingham's Department of Botany in 1975, Cocking and his team had fused a member of the grape family (*Parthenocissus tricuspidata*) with a petunia (*Petunia hybrida*). The chromosome segregation of the fused plant indicated "the possible limitations of somatic hybridization between distantly related plant species."[92] Despite the loss of one set of chromosomes, some hybrid cells did survive. These contained a mixture of cytoplasm from both species but with the nucleus of the *Parthenocissus*.[93] Cytoplasmic hybridization therefore seemed to provide a means of overcoming chromosome segregation to transfer desirable characteristics between distinct plant species.

In a 1983 paper in *Science*, a team from the Plant Pathology Department at Kansas State University described the production of four somatic hybrids following cell fusion between Russet Burbank potatoes and Rutgers and Nova tomato cultivars. Chromosome counts indicated that chromosome segregation had not been complete: some of these regenerated "pomato" plants showed "a more tomato-like morphology . . . including more intense red pigmentation, more pointed terminal leaflets, and more extensive leaf serration." The plant pathologists concluded that small chromosome segments might have survived, offering the possibility of using cell fusion for "introducing genes from unconventional sources." The experiments conducted at Nottingham and Kansas State seemed to indicate that partial somatic hybridization, or cytoplasmic hybridization, offered a means of overcoming chromosome segregation to combine completely different plant species. Furthermore, these techniques could prove valuable, as "directed transformation with cloned genes," or recombinant DNA technology, was still relatively unsophisticated.[94] However, recombinant DNA technology was always improving and was admittedly a more precise way of introducing new genes.

"Research on protoplast fusion," announced David A. Evans, associate scientific director at the DNA Plant Technology Corporation, in 1983, "has been increasingly focused on the transfer of organelle-encoded traits." While this may have been the case, the biological mechanism underpinning cytoplasmic hybridization was complex. It was only by the early years of the 1980s that a consensus was beginning to emerge on how cytoplasmic genetics actually worked.[95] Theoretical complexity and disagreements did not discourage plant scientists and geneticists from attempting to apply cytoplasmic hybridization to agriculture. Several laboratories, including those at the University of Nottingham and Kansas State University, had attempted to transfer cytoplasm via cell fusion. Evans recognized that a number of "agriculturally useful traits are cytoplasmically encoded," including "male sterility," which is important in the hybridization of wheat, "and certain herbicide and disease resistant factors."[96] However, like Cocking before him, Evans stressed the need for those working on cell fusion to talk with their counterparts in "plant genetics and plant breeding to encourage interchange of biotechnology objectives."[97] If cell fusion did not meet the needs of agriculture, the hope of developing practical products from the technology was remote.

The need to transform both somatic and cytoplasmic hybridization into agricultural biotechnologies was also at the forefront of Cocking's mind in 1987. By this time, cell fusion—involving the transfer of nuclear and cytoplasmic encoded genes—had been conducted in major crops, such as rice, to induce salinity tolerance and disease resistance. Yet despite such advances in somatic hybridization, scientific interest in recombinant DNA technology had surged instead. The 1984 international symposium on genetic manipulation in crops had shown that more and more researchers considered recombinant DNA technology to be a plant-breeding technology superior to cell fusion. Writing in *Science* with coauthor Michael R. Davey of the School of Biosciences at the University of Nottingham, Cocking considered combining the two technologies. Cell fusion could be quickly and practically applied to agricultural crops—as had been accomplished with rice varieties in 1986—as "non-recombinant DNA somatic cell fusion procedures . . . do not rely for their implementation on a detailed knowledge of the genes involved."[98] "Somatic hybridization and genetic transformation" were "two radically different approaches to manipulation of plant genomes." Following a cell fusion event, whether involving nuclear or cytoplasmic transfer, resulting plants could simply be screened and selected based on their characteristics in a manner familiar to traditional breeders. Recombinant DNA technology, according to Cocking, was "dependent upon relating genotype to phenotype in a tangible way so as to ascertain what biochemical and developmental activity is controlled or modulated by a DNA sequence." Establishing this relationship from genotype to phenotype was no easy feat, but could be avoided entirely by using cell fusion.[99]

Cocking was not entirely combative in his comparison of plant cell fusion with recombinant DNA technology. By the early 1980s it had become apparent that protoplast research and recombinant DNA technology could complement each other. In the 1981 issue of the *Philosophical Transactions of the Royal Society*, Cocking explained how protoplasts could aid those struggling to insert sections of foreign DNA into crop plants using *Agrobacterium tumefaciens*, which "naturally manages to transfer, maintain and express its prokaryotic DNA in plant cells." By inserting the bacteria into protoplasts, the number of transformed plants available to researchers could be significantly increased.[100] In a similar manner, Davey and Cocking claimed in 1987 that "gene transfer in cereals" using somatic or cytoplasmic hybridization could benefit from "close integration of cell culture and molecular approaches."[101] Details on the shape of this collaboration, however, were not forthcoming.

Cytoplasmic hybridization was pursued during the 1980s as the limits of somatic hybridization became apparent. Chromosome segregation meant that a far smaller range of crop varieties could be crossed with each other to form fertile somatic hybrids than was once believed. Moreover, plant physiologists and geneticists realized how much chromosome segregation limited the potential of somatic hybridization, just as the potential of recombinant DNA technology to bypass these same limits became apparent. Cytoplasmic hybridization therefore appeared to offer advocates of cell fusion the timely opportunity to cross crop plants from diverse species. Yet throughout the 1980s, cytoplasmic hybridizers were seemingly forced to justify the practicality of their biotechnology by appealing to its advantages over recombinant DNA technology. There were some successes in the production of new crop varieties and some attempts to integrate cell fusion with recombinant DNA technology. Ultimately, though, these efforts were not enough for plant cell fusion to regain its place as the premier form of modern plant biotechnology.

For approximately twenty years, encompassing the 1970s and 1980s, somatic hybridization did not create commercial plant breeds. The technique was instead plagued by slow and periodic development. Its complexity and unpredictable nature were also uninviting to its potential users; namely, plant breeders. Cocking was aware of this problem, urging "protoplast workers" to engage in "a continuing dialogue with breeders."[102] Yet there are recognized benefits to studying a seemingly failed innovation. Despite its commercial failure, somatic hybridization is revelatory of both the ambitions of plant physiology and the wider collaborative attempts to exploit the plasticity of living things on the cellular level since the 1960s. Like other twentieth-century attempts to transform plant breeding, somatic hybridization is an example of a technique that has been largely "lost to the history of biotechnology, and yet constitute[s] an important component of that history."[103] Although somatic hybridization had moved much further along the commercial pipeline than graft hybridization ever did, both techniques were dogged by similar issues: the technical problems of raising fragile plants, identifying hybrid traits, and competing with rivals.

A Hybrid Analogy

Given that somatic and graft hybridizers both pursued the aim of breaking the species barrier through the merging of cellular material, we would expect to find some crossover between the two techniques. This convergence took different forms at different times. During the early twentieth century, we find Hans Winkler turning to cell fusion after the loss of his prized graft hybrid. In the postwar era we see the Brookhaven National Laboratory team turning to grafting to nurture their fledgling somatic hybrid. In the wake of the Lysenko affair, the team explicitly stated that only cell fusion could result in the exchange of genes between plants. Grafting had nothing to do with it. If we take a step back from this denial, however, we can spot numerous similarities between graft hybridization and cell fusion. The history of the latter is very much like the history of graft hybridization in miniature. Both operated largely outside of mainstream genetics, with cell fusion pioneered by an odd mix of virologists and plant physiologists. Practitioners of both techniques struggled to construct a history for themselves, as we might expect from endeavors driven by experiment over theory. "If techniques tend to play second fiddle to ideas in the annals composed by the historians and

CELL FUSION

philosophers of science," wrote the 1969 contributor to *Nature*, "that is perhaps because they burst on the world with less éclat than the grand theories such as sea-floor spreading or the Watson-Crick hypothesis."[104]

The struggle of cell fusion to contribute to agriculture arose from the problems that also faced the graft hybridizers we have thus far encountered. Cell fusion was difficult to perform and had a low success rate. Mixing the contents of cells, even when successful, gave an unpredictable outcome. Waste and failure were unavoidable, an issue encountered by almost every graft hybridizer from Charles Darwin to Anne McLaren. Definitively identifying plants as hybrids also posed a challenge for graft hybridizers, while advocates of cell fusion—even with more sophisticated methods at their disposal—had to contend with chromosome shedding. Finally, both techniques faced strong competition in the agricultural sphere. Mendelian crosses and sexual hybridization proved far more reliable than grafting, while cell fusion was largely abandoned in favor of recombinant DNA technology. The case of cell fusion demonstrates how a technique can falter in the long and tortuous route from laboratory innovation to practical product. Even with sufficient financial backing and enthusiasm from multiple scientific disciplines and the public, somatic hybridization did not make the transition to agriculture until late in its career. Both graft hybridization and cell fusion lacked "robustness," in that they were technically difficult to pull off and thus off-putting to researchers. Cell fusion also faced difficulties integrating itself into existing industrial systems, with demand for the enzymes used to create protoplasts outstripping supply into the 1970s.

On a broader scale, cell fusion also forces us to reconsider what we define as "biotechnology" or "genetic engineering." There is an ongoing debate in historical circles over the meaning and scope of what we term *biotechnology*, a perplexity reflected in current definitions released by industry and government. This was characterized as a divide between "ancients" and "moderns" in a 1986 edition of the French biotechnology journal *Biofutur*.[105] A modern view of biotechnology only begins with the discovery of the structure of DNA in 1953 and subsequent developments in molecular biology. Advocates of the ancient view embraced a much wider conception of biotechnology, envisioning a three-stage history that moved from Egyptian and Babylonian brewing to Pasteurian-informed "rational fermentation," and finally to "genetically based molecular biology."[106] Somatic hybrids do not fit into any of these conceptions or categories. Plant cell fusion is certainly not a form of brewing or rational fermentation, nor does it manipulate organisms on a

185

genetic level. Graft hybrids also confound these categories, a problem we will turn to in the concluding chapter of this book. It is important to note, too, that the dominance of recombinant DNA technology over cell fusion was not inevitable. At the right time, a surplus of cellulase enzyme, the support of geneticists, or mishaps in the development of recombinant DNA for agricultural use might all have shifted the balance of history in favor of cell fusion. Graft hybrids also fell prey to the vagaries of historical contingency.

CONCLUSION

A Modern Rediscovery

—◯

IF THE DAWN OF THE TWENTIETH CENTURY WAS MARKED BY THE
famous "rediscovery" of Mendelian laws, then the dawn of the twenty-first
was marked by a rather more muted rediscovery. In 2009 Sandra Stegemann
and Ralph Bock, biologists at the Max Planck Institute of Molecular Plant
Physiology in Potsdam, Germany, overturned assumptions that genetic material could not be exchanged between two grafted plants. By generating
two varieties of transgenic tobacco with genetic markers, they were able to
show that chloroplast genes moved across the graft junction.[1] These modern rediscoverers were careful to note that this genetic exchange did "not
lend support to the tenet of Lysenkoism that 'graft hybridization' would be
analogous to sexual hybridization." In a way, they downplayed their findings,
observing that gene transfer between the tobacco plants had occurred only
in the area immediate to the graft junction, restricting inherited changes
to plant shoots emerging directly from the junction. This phenomenon,
however, would almost certainly have been taken by their botanist forebears
as evidence of graft hybridization. Stegemann and Bock concluded that any
future reports of heritable changes in plants following grafting did "warrant
detailed molecular investigation."[2]

A few years later, further work on grafted plants by researchers based at
the Max Planck Institute and the Department of Molecular Biology at the

John Paul II Catholic University of Lublin brought more intriguing findings. In 2014 the teams conducted experiments grafting transgenic tobacco plants. They found that "nuclear gene transfer across the graft junction had occurred."[3] When the cells of plants arising from a grafted parent were examined, they were found to be larger than usual. While the cells of the parent tobacco plants typically contained forty-eight chromosomes, some of their offspring contained up to ninety-six. Chromosome doubling had occurred through grafting. "We have demonstrated," declared the study's authors, "that grafting results in the transfer of entire nuclear genomes between species." They also theorized that, since grafting can occur naturally, it presented biologists with "a potential asexual mechanism of speciation." Artificial grafting could also be used to create new crop species through polyploidy, thereby conferring "the superior properties of modern crops over their diploid progenitor species. This has significant potential in breeding and agricultural biotechnology."[4]

Only a few years later, another team of scientists, this time at Rutgers, the State University of New Jersey, found that mitochondria could also pass between grafted tobacco plants. "Graft transmission of mitochondria," reported the group, "provides a mechanism for horizontal transfer of entire genomes via mitochondrial fusion."[5] The evolution of parasitic plants, which latch onto their hosts, may have benefited most from this horizontal gene transfer. Mitochondrial movement between grafted plants could also, they suggested, have applications in biotechnology. For instance, it could be used to induce cytoplasmic male sterility, where plants cannot produce functional pollen, to facilitate the production of hybrid seed.[6] With molecular-level analysis, biologists had found that all three genomes present in plants could pass across the graft junction. Most importantly, nuclear cell fusion—once considered the most likely mechanism for the creation of graft hybrids—could occur. The graft hybrid was back in town. Almost overnight, it had gone from a scientific pariah to a novel evolutionary mechanism and a player in our biotechnological future. Contemporary scientists are, in effect, rewriting the past, present, and future of the graft hybrid.

The Century of the Graft

Beyond the specific arguments I have made in this book, the story of the graft hybrid holds some general insights. One is the role of chance and contingency in the history of science, which played an outsized role in the failure of the graft hybrid to be integrated into twentieth-century heredity. At key moments in the history of the graft hybrid, compelling evidence has been destroyed or lost. Shortly after Charles Claude Guthrie announced the results of his transplantation and breeding experiments in chickens, the birds succumbed to a disease on his family farm. Disheartened, Guthrie did not attempt to replicate the experiment. Hans Winkler also possessed living evidence of graft hybridization in the form of the *Solanum darwinianum*, but this plant was lost during the First World War. Later in the century, other promising experiments proved difficult to replicate. Why were graft hybrids so vulnerable to the vagaries of chance and circumstance? The answer lies in the organisms themselves. Part of the success of genetics came down to its practitioners' ability to demonstrate their theories using experimental organisms, usually mammals, some of which later acquired the label "model organism."[7] Graft hybrids, in contrast, were largely created using plants, which can be extremely complex in terms of heredity—a lesson learned by both Mendelians and the followers of Hugo de Vries in the early twentieth century.[8] Nor did graft hybridizers ever make the leap to the classic "model organisms" of genetics, such as fruit flies or laboratory rats.

Given that graft hybrids were often victims of circumstance, it is possible to imagine that they might have become a part of mainstream biology. Historians of science have long called for the reinstatement of alternative theories of heredity in the historical canon. However, one criticism leveled at this approach is that the experimental results of these alternate theories of heredity were ambiguous or simply nonexistent in comparison to those of classical genetics.[9] Graft hybrids certainly suffered from this shortcoming, in the sense that they presented biologists with a phenomenon that was difficult to interpret. As the loss of graft hybrids across the twentieth century shows, from Guthrie's chickens to Paul Kammerer's salamanders, one-off demonstrations of graft hybridization were all too often fragile and vulnerable. We cannot, however, simply explain away graft hybrids by suggesting that the death of living examples of the technique determined its failure. A recent study of another Mendelian controversy suggests that evidence of just how close debates over the future direction of heredity came can be found in "the manner in which Weldonian [in reference to W. F. R. Weldon, a leading biometrician] emphases did not so much disappear from genetics as take up permanent residence on the conceptual sidelines [of biology]."[10] Graft hybrids also occupied this space, with only the contingencies of experimental ambiguity and the loss of specimens barring them from mainstream biology. The twentieth century was not as kind to the graft as it was to the gene. Yet the graft hybrid was a survivor. If it had succeeded in becoming an established scientific fact prior to the twenty-first century, what might this have done to biology?

Alternate Histories of Heredity

In the opening of her classic text on the "century of the gene," Evelyn Fox Keller observed that when the term *gene* was introduced by Danish botanist Wilhelm Johannsen in 1909, nobody really knew what it was. In 1933 American geneticist Thomas Hunt Morgan would admit that his discipline could not agree on what genes were, or even if they were real or fictitious. Despite this shortcoming, however, genes had by then effectively "become incontrovertibly real, material entities—the biological analogue of the molecules and atoms of physical science."[11] All this tells us is that genetics was a painstakingly constructed system, with its characteristics contingent upon the outcome of "epistemic ruptures and conceptual setbacks, political controversies and struggles for authority among diverse modes of knowledge, and national styles of thought."[12] Equally, the embrace of "hard heredity," which separated the gene from its surrounding environment and somatic cells, was not inevitable. At various points throughout the twentieth century, opportunities arose for scientists to debate theories or experiments that did not necessarily fit with their understanding of classical genetics. As both a concept and an experimental program, the graft hybrid was therefore well placed to intervene in discussions of heredity.

One such opportunity for graft hybridizers arose in the first decade of the twentieth century. During the late nineteenth century, the study of the cell had led to great excitement among biologists that a mechanistic explanation of heredity might be uncovered. This focus on cytology had ultimately reduced the "power of environmental factors in explaining how variations and evolution occurred."[13] Prior to the First World War, cytological research in Britain and Germany had indicated a physiological basis for the exchange of hereditary material across plant grafts. Counting chromosomes became the nexus for debate over whether Winkler had succeeded in producing a true graft hybrid. The minuscule threads connecting cells in grafted plants, uncovered by Cambridge botanist Margaret Hume in 1913, were reinterpreted by some graft hybridizers as a means by which genes could pass between plants. The very techniques that had once restricted heredity to the sex cells could effectively be deployed in reverse, extending inheritance to the bodies of grafted plants and beyond. We can also argue that grafting experiments might have provided an alternative theory of heredity if not for a series of missed opportunities. If only Guthrie's chickens had survived his family farm and produced more offspring . . . If only Winkler had kept the *Solanum darwinianum* alive and produced his promised book on graft hybrids and chimeras . . . We could keep expanding this list, including more "what ifs" relating to the animal experiments of Kammerer, or the hunt for graft hybrids in botanic gardens by Frederick Ernest Weiss during the 1930s.

Yet citing these missed opportunities is not enough. To produce a genuine alternative to classical genetics, we would need to find a coherent system of heredity capable of standing in its place.[14] A lone graft hybrid would not have provided this, although its inclusion in biological thought might have altered our thinking about heredity in interesting ways. Take, for instance, the division of biological thought into two classes by Ernst Mayr in 1982. Mayr naturally focused on his favored class, "evolutionary biology," in which he included such luminaries as Charles Darwin, August Weismann, and Hermann Joseph Muller. He excluded "physiological biology" altogether. This distinction has subsequently been criticized on the grounds that characters such as Darwin quite happily practiced natural history and physiology without observing any such divide.[15] Had the graft hybrid made its way into twentieth-century biology, however, Mayr's division would have been impossible to make. Heredity could not be understood without physiology, while physiology could not be studied without considering its impact on inheritance. Despite the hostile attitudes of some geneticists to graft hybridization, it is unlikely that the acceptance of graft hybrids would have excluded genetics completely. We have encountered several examples of Mendelian genetics and graft hybridization coexisting. Guthrie and Winkler thought that their graft hybrids represented a new heredity mechanism, rather than overthrowing the subject altogether. Kammerer conducted Mendelian crosses on salamanders, while Ivan Vladimirovich Michurin did not "deny the merit" of the "Mendelian law."[16] When considered against these examples, Trofim Lysenko's outright denial of genetics was an extreme deviation from the norm. The gene could have remained intact, but heredity would have been extended to incorporate the influence of the somatic cells. It is not clear whether this extension would have included the Lamarckian influence of the environment on heredity, as feared by many geneticists.

A more powerful case for a counterfactual history of graft hybrids can be found in the world of plant and animal breeding. As I have argued, techniques were embraced by breeders based on their utility. Theoretical coherence was of secondary importance. The activities of members of agricultural and plant breeding bodies in Britain provide us with some clues that grafting was taken seriously by this community during the Lysenko affair. The ability of graft hybrids to bypass the limits on sexual hybridization undoubtedly made them extremely appealing to breeders.[17] Anne McLaren would seize upon this interest by promoting the idea that grafting or transplantation could invoke "hybrid vigor" in plants or animals. Through the work of Joseph Needham—in particular, his multivolume *Science and Civilisation in China*—the influence of the graft hybrid as a horticultural tool has made its presence felt in the discipline of history. By the 1930s the Cambridge-educated biochemist had already turned his attention to history, writing books on the history of embryology and the Leveler movement of the English Civil War. The latter was published under a pseudonym, as the Marxist Needham feared that his politics would lead to a backlash from his Cambridge colleagues.[18] Needham's first visit to China in 1943 confirmed his belief that the "scientific heritage" of China would merge with its "forthcoming socialist transformation" to produce "the future political and scientific front line for all humankind."[19] Needham subsequently began writing *Science and Civilisation in China*, the first volume of which appeared in 1954.

Science and Civilisation in China covered some twenty-five centuries and subjects as diverse as astronomy and metallurgy. Its volumes on "biological technology" were not published until the mid-1980s, with assistance from Needham's collaborators. Yet Needham himself had begun investigating plant grafting some thirty years earlier. In the 1950s and 1960s he reached out to various scholars for information on grafting in Europe and China. Michael Joseph Hagerty, a Chinese translator at the Office of Physiology and Breeding Investigations in the United States, provided Needham with references to grafting in the 1726 "Thu Shu Chi Chhêng," an imperial encyclopedia.[20] Cambridge classicist William Palmer gave him a list of references to grafting in Virgil's *Georgics*.[21] Needham's own notes from this period indicate that he was engaging with Chinese texts whose authors claimed to be able to change the "fixed characteristics" of nature through grafting; for example, transforming pink flowers into purple ones.[22] In the volume of *Science and Civilisation in China* on botany, Needham would repeat these claims, stating that "people competed to produce new varieties by cultivation and grafting" and that "many exquisite ones appeared."[23] Needham even cited William Neilson Jones's 1934 book on graft hybrids.[24]

What happened shortly before the 1986 publication of the volume of *Science and Civilisation in China* on botany is even more intriguing than Needham's previous encounters with grafting. In 1984 Victor Meally, an Irish mathematician who had provided comments on earlier volumes, sent Needham a set of newspaper clippings concerning the Lysenko affair. One article by the biologist Conrad Hal Waddington, published in 1949, stated that Lysenko's claims were not easy to accept. However, "the so-called 'graft hybridization,'" wrote Waddington, "is perhaps the less difficult." The graft hybrid problem, he acknowledged, had a long history. It was possible that Lysenko was "on to something" regarding the existence of graft hybrids. However, argued Waddington, graft hybridization would not overturn classical genetics but might provide "an addition, even quite an important addition, to it." The second clipping was a transcript from J. B. S. Haldane's infamous 1948 radio interview, in which Haldane claimed that "one can sometimes change the character of a living creature by grafting it, and the changed character is sometimes inherited." Needham responded by politely thanking Meally for the "interesting" newspaper clippings, but gave no indication that he believed their claims or would consider including them in *Science and Civilisation in China*.[25] If his response had been different, graft hybrids could have appeared in a major historiographical work much earlier, incorporating a Chinese-oriented perspective with a sprinkling of Marxism.[26]

Needham's close encounter with the graft hybrid reinforces the sense that the phenomenon was omnipresent on the fringes of biology, ready to intervene at any moment to challenge genetic explanations of heredity or transform the practice of breeding. For the first half of the twentieth century, the graft hybrid was part of an ongoing discourse on the nature of heredity. With the rise of Lysenko in the second half of the twentieth century, it became increasingly difficult for Western biologists to work on non-Mendelian inheritance. Their politics, as well as their science, had become suspect. In the West, this ideological policing "limited scientific heresies," yet "also impeded the creativity inherent in them."[27] Across the entirety of the twentieth century, biologists did not underestimate the importance of graft hybrids. The geneticist Richard Goldschmidt remarked in 1914 that the "mystery" surrounding graft hybrids had been a "bone of contention among horticulturalists for several centuries."[28] Three decades later, P. S. Hudson and R. H. Richens, of the Cambridge-based Imperial Bureau of Plant Breeding and Genetics, stated that the question of whether graft hybridization could occur was one of biology's longest-standing problems.[29] Historians, too, have begun to recognize its importance. This recognition is particularly timely given the twenty-first-century resurrection of the graft hybrid.

A Biotech Future?

"Genetic engineering is the single most important development in biology since Charles Darwin's exposition on the origin of species by means of natural selection in 1859," declared Robert Pickard, a neurobiologist and scientific advisor to the British government, in 2001. At the dawn of the twenty-first century, Pickard understood that "evolution by natural selection and the modification of genotype to achieve phenotypic goals [changes to the body of the organism] have evoked great controversy."[30] The graft hybrid, too, has seen its fair share of controversy. It is analogous to biotechnology in more meaningful ways as well. *Contemporary biotechnology* is often defined as the engineering of life for commercial or industrial purposes through genetic modification. Yet biotechnology has also been defined as an archaic practice, encompassing everything from ancient breeding and brewing to Pasteurian "rational fermentation" and "genetically based molecular biology."[31] *Plant biotechnology* has been harnessed by historians as a useful term by which to refer to several different techniques in the history of horticulture, including grafting and cloning.[32] Under this broad definition, early twentieth-century farms, factories, and laboratories were sites of biotechnology. Microbes and molds drove fermentation-based industries and sewer systems. Jacques Loeb, who developed artificial parthenogenesis in 1899, sought to mimic the efficiency and rationality of engineering in biology.[33] In 1920 a Bureau of Biotechnology was established in Leeds, England, to explore the use of microorganisms for brewing, sterilization, and the control of fungal and insect pests.[34]

Given this history, we should therefore have few qualms about labeling graft hybridization as a biotechnology. Besides providing us with a useful label, incorporating the graft hybrid into the history of biotechnology helps us to think of this history not as a series of linear developments but as a diverse set of overlapping—and often unsuccessful—techniques. Acknowledging this diversity changes traditional narratives of who has participated in the development and application of biotechnology. These new participants might include brewers, desktop experimenters, and horticulturalists.[35] In this respect, the graft hybrid was quite unlike Mendelian genetics—which, as we saw in the introduction to this book, readily merged with a capitalist drive toward accounting and standardization in early twentieth-century agriculture. Mendelism and hybridization were accompanied by the wresting of plant

FIGURE 14. Multiple grafts. Now that scientists have demonstrated that genes can move across the graft junction, graft hybridization is poised to enter the world of commercialized genetic engineering. Complex grafts across different species, as depicted in this image from the early twentieth century, may become more and more common. Lucien Louis Daniel, "Sur la réussite, le développement, la durée et la production des greffes (suite)," *Revue bretonne de botanique pure et appliquée* 6, no. 4 (1911), plate VII.

breeding away from farmers and into the hands of "expert" breeders or geneticists. Grafting, on the other hand, could be done by anyone at any time. It was, in some respects, still a "subversive" art in the twentieth century. If new plant hybrids could be made by grafting, the concentration of hybridization in the hands of private companies and the power of intellectual property laws in biology could effectively be bypassed (for an example of an easily made plant which combines several species through grafting, see fig. 14).[36]

The modern rediscovery of the graft hybrid has seen an explicit attempt to cast off its history and Cold War legacy in favor of integrating it into the world of commercial biotechnology. Just prior to its triple rediscovery, Yongsheng Liu praised the anticapitalist nature of the graft hybrid. Graft hybridization was "a simple and powerful means of plant breeding" that could compete with such techniques as recombinant DNA technology.[37] This sentiment is not echoed elsewhere. Stegemann, Bock, and their colleagues have suggested that the movement of chloroplasts across the graft junction "opens possibilities in plant breeding and also facilitates genetic-engineering approaches in species that are recalcitrant to plastid genome transformation."[38] Similarly, the researchers from the Max Planck Institute and the John Paul II Catholic University of Lublin portrayed "grafting-mediated genome transfer" as "a new tool for crop improvement." It could, for instance, double the chromosome count of crops, thereby conferring "the superior properties of modern crops over their diploid progenitor species [which] has significant potential in breeding and agricultural biotechnology."[39] Finally, the Rutgers University researchers argued for the application of grafting in biotechnology. It was "a versatile approach" that could be used to move organelles between plants or create "nuclear hybrids."[40] Rather than marking a new direction for plant breeding, graft hybridization would be used to support or enhance existing systems.

"We have been accidentally genetically engineering plants—and eating GMOs—for millennia," announced Michael Le Page, a reporter for *New Scientist*, in March 2016. Le Page was referring to recent studies on the movement of genes between grafted plants. He noted that graft hybridization "could provide plant breeders with new tools to create novel traits and crops. Bock is already trying to use grafting to create new species, such as a tomato-chilli mix." The transfer of mitochondria across the graft junction also offered a promising development for plant breeders. Grafting now "offers a way to transfer traits encoded by mitochondrial genes, such as male sterility, to plants that lack them." Le Page clearly envisioned a future where graft hybridization would play a prominent role in plant breeding. His interest in the technique was not only economic. He also notes in his article that "it is highly likely that some of the plants we eat were created by this kind of unintentional genetic engineering by farmers." This finding had contemporary relevance. "The idea that we have been unintentionally modifying plants by grafting," wrote Le Page, "will not be welcome to those who like to claim that grafting is very different to genetic modification."[41] In other words, for as long as humans have been grafting, we have been moving genes between plants to create transgenic organisms.

Why does this reinterpretation of the deep history of plant breeding matter? The movement of genes across crop varieties and species using genetic modification is seen by many members of the public as undesirable, or unnatural.[42] By equating grafting with genetic modification, advocates of the latter can make new claims for its "naturalness" and antiquity. This comparison is not without merit. Take the media and public interest in cloning following the birth of Dolly the sheep at the Roslin Institute in 1996. Given that the American plant physiologist Herbert John Webber had introduced the concept of cloning back in 1903, nothing about Dolly "was the result of fundamentally new science."[43] Rather than reciting the numerous twenti-eth-century developments in cloning that made Dolly possible, media atten-tion focused instead on the rather technical detail that an adult somatic cell, rather than an embryonic one, was used in the cloning process. Subsequent stories focused on the possibility of cloning adult humans and all the thorny ethical issues that would arise in such a scenario. An opportunity therefore exists for historians of biology to "illuminate public discussion and media presentation" of the biological sciences—in the case of cloning, by demon-strating that it "is not radically new science."[44] In the case of the graft hybrid and genetic modification, it is not at all clear that such a history is desirable. Crude attempts to compare genetic modification to earlier plant breeding techniques have been a standard part of the arsenal of biotech firms such as Monsanto since the 1980s. Sketching out the long history of biological tech-niques can simply play into corporate narratives without meaningfully en-gaging with the social and political issues surrounding new biotechnologies.

The history of the graft hybrid suggests that it will not readily become part of plant breeding without attracting controversy. In the West, the in-tegration of Mendelian genetics and breeding was told through the story of hybrid corn (maize). Between 1910 and 1935, traditionally bred—either by deliberate inbreeding or open pollination—American corn varieties were gradually replaced by their hybrid counterparts.[45] In the Great Depression, the principal concern of the American public was to obtain "ample and affordable supplies of food, clothing and shelter," which indicates why this "strange new creation of science" was widely accepted with little or no public outcry.[46] In a 1958 visit to the United States, Soviet premier Nikita Khrushchev stopped to visit the hybrid-seed-producing farm of Roswell Garst. A triumphal moment came when a jovial Khrushchev held aloft an ear of Garst's hybrid corn in front of a crowd of reporters. As a photo op-portunity, the moment was hard to beat. Khrushchev's tacit endorsement of hybrid corn held symbolic connotations: hybrid corn was living proof of the truth and utility of Western genetics in plant breeding and confirmed the bankruptcy of Lysenko's biology.[47] Yet it remains a hotly contested point that "agricultural reality," in the form of hybrid corn, "crushed a would-be rival to genic biology."[48]

Graft hybrids may one day have their hybrid corn moment, with a living and economically useful specimen finally demonstrating their "agricultural reality." What would our reaction be to such a moment? The hybrid corn story cemented the sense that the "values" of Mendelian genetics aligned with those of 1950s America. At the time of writing, it is not clear what form a graft hybrid "success" story might take or who will tell it. Unlike hybrid corn, graft hybrids are claimed by both Marxist breeders and Western molecular biologists. They may be a kind of biotechnology, but it is not a kind that fits easily into the commercialized world of contemporary genetic modification. After all, if Le Page is correct, anyone can potentially create a graft hybrid. If they can occur so easily and so randomly, regulating their creation or their ownership will be a near-impossible task. Predicting the place of the graft hybrid in the biotech age is an equally difficult challenge. It is at least testament to the resilience of the graft hybrid that an ancient plant-breeding technique, first defined by Darwin in 1868, has now been called upon to influence the future of genetically modified crops.

NOTES

Introduction

1. Frank and Chitwood, "Plant Chimeras," 41–53.
2. Ragionieri, "Origin of the Florentine Bizzarria," 527–28.
3. Savoia, "Nature or Artifice?," 67–86.
4. Weiss, "Problem of Graft Hybrids and Chimaeras," 264.
5. For a recent take on the history of cloning, see Crowe, *Forgotten Clones*.
6. Campbell, *Botanist and the Vintner*; and Mudge et al., "History of Grafting," 437–93.
7. Churchill, *August Weismann*, 527.
8. Müller-Wille and Rheinberger, *Cultural History of Heredity*, 8.
9. Darwin, *Variation of Animals and Plants under Domestication*, 390.
10. Wardy, "Mysterious Aristotelian Olive," 69–91.
11. Pease, "Notes on Ancient Grafting," 66–76. For a similar piece, see Roberts, "Theoretical Aspects of Graftage," 423–63.
12. Lowe, "Symbolic Value of Grafting in Ancient Rome," 482.
13. "Correspondent," 35.
14. Risso and Poiteau, *Histoire naturelle des orangers*, 47.
15. Risso and Poiteau, 107–8.
16. Darwin, *Variation of Animals and Plants under Domestication*, 387.
17. For an account in English, see Poiteau, "Remarks on the *Cýtisus Adàmi*," 58–61.
18. Herbert, "On the Singular Origin of the Purple Laburnum," 289–90. Herbert came up with his own "recipe" for producing graft hybrids, advising his readers to lacerate the edges of stock and scion to completely intermix the tissues of both plants. See Herbert, "Further Remarks on the Cytisus Adami," 381–82.
19. Lidwell-Durnin, "Production of a Physiological Puzzle," 14. Lidwell-Durnin notes that Darwin's intervention occurred only after seventeen years of debate over the plant. The importance of *Cytisus adami* to Darwin's views on heredity is covered in Browne, *Charles Darwin*; and Olby, *Origins of Mendelism*.
20. Explanation from Geison, "Darwin and Heredity," 375–411.
21. Darwin, *Variation of Animals and Plants under Domestication*, 397. On the reception of pangenesis, see Holterhoff, "History and Reception of Charles Darwin's Hypothesis," 661–95.
22. Friedrich Hildebrand to Charles Darwin, 2 January 1868, DCP-LETT-5774, Charles Darwin Correspondence. To my knowledge, Hildebrand's exchange with Darwin on graft hybrids has not yet been discussed in the historical literature.
23. Hildebrand to Darwin, 4 November 1868, DCP-LETT-6448, Darwin Correspondence.
24. Darwin to Hildebrand, 14 November 1868, DCP-LETT-6459F, Darwin Correspondence.
25. Schwartz, "George John Romanes's Defense of Darwinism," 289–93. Following Darwin's death in 1882, Romanes reverted to animal experiments. His efforts to graft hybridize animals were as unsuccessful as his efforts with plants.

26. Galton, "Experiments in Pangenesis," 403–4. Darwin protested these findings on the grounds that he had never claimed gemmules were passed on through blood. Galton backed down.

27. McIntosh, *Book of the Garden*, 326–27.

28. Corsi, "Jean-Baptiste Lamarck," 18. On the reception of Lamarck, see Galera, "Impact of Lamarck's Theory of Evolution," 53–70.

29. Burkhardt, "Lamarck, Cuvier, and Darwin on Animal Behavior," 33–34.

30. August Weismann would later define Lamarckism as a mechanism of inheritance of environmental effects. On the "strawman" qualities of this definition, see Gliboff, "Golden Age of Lamarckism," 49.

31. Müller-Wille and Rheinberger, *Cultural History of Heredity*, 81.

32. Müller-Wille and Rheinberger, 81. Following Boveri's experiment, cell research abandoned its predominantly descriptive orientation and entered an experimental phase.

33. Churchill, *August Weismann*, 197.

34. Sapp, *Beyond the Gene*, 4.

35. Bowler, *Eclipse of Darwinism*, 65.

36. Bowler, 62–63.

37. A brief summary of the various disciplines involved in the study of heredity (including genetics, embryology, cytology, and natural history) can be found in Sapp, "Struggle for Authority," 316. On Loeb, see Pauly, *Controlling Life*.

38. Sapp, *Beyond the Gene*, 55–56.

39. Lönnig and Saedler, "Baur, Erwin," 199–203.

40. Sapp, *Beyond the Gene*, 56.

41. Sapp, 232. Argument from Witkowski, review of *Beyond the Gene*, 233–34.

42. Churchill, *August Weismann*, 532.

43. Churchill, 546.

44. Winther, "August Weismann on Germ-Plasm Variation," 550.

45. Campos, "Genetics without Genes," 247.

46. Campos, 254.

47. Joravsky, *Soviet Marxism and Natural Science*, 299.

48. Müller-Wille and Parolini, "Punnett Squares and Hybrid Crosses," 149–65.

49. Example from Bateson, *Mendel's Principles of Heredity*, 15.

50. Radick, *Disputed Inheritance*, 4.

51. Wood and Orel, *Genetic Prehistory in Selective Breeding*, 271–74. A similar argument is given in van Dijk, Weissing, and Ellis, "How Mendel's Interest in Inheritance Grew," 347–55.

52. Radick, *Disputed Inheritance*, 301–2.

53. On what Mendel discovered and the problems of the 1900 rediscovery, see the classic texts: Brannigan, "Reification of Mendel," 423–54; and Olby, "Mendel No Mendelian?," 53–72. On de Vries, see Stamhuis, "Why the Rediscoverer," 29–49. The status of Mendel as the founder of genetics was cemented during the Cold War. Wolfe, "Cold War Context of the Golden Jubilee," 389–414.

54. For an overview of the debate, see Harwood, "Did Mendelism Transform Plant Breeding?," 345–70.

55. Fitzgerald, *Every Farm a Factory*.

56. Thurtle, *Emergence of Genetic Rationality*.

57. Allen, "Origins of the Classical Gene Concept," 8–39. Early Mendelians also compared their science with synthetic chemistry. See Müller-Wille and Rheinberger, *Cultural History of Heredity*, 137.

58. Elina, Heim, and Roll-Hansen, "Plant Breeding on the Front," 161–62.
59. This realization was reflected in farm practice. Fitzgerald, "Farmers Deskilled," 324–43.
60. Bonneuil, "Mendelism, Plant Breeding and Experimental Cultures," 296–97.
61. Charnley, "Experiments in Empire-Building," 292–300.
62. Endersby, "Mutant Utopias," 481.
63. Allen, "Hugo de Vries," 55–87. The economic ambitions of mutation theory lived on through mutation breeding. See Curry, *Evolution Made to Order*.
64. Kohler, *Lords of the Fly*.
65. Burian and Zallen, "Genes," 438.
66. Sapp, *Beyond the Gene*, 3.
67. Gliboff, "Case of Paul Kammerer," 525–63.
68. Cook, "Bacon Predicted Triumphs of Plant Breeding," 162.
69. Characterization of Lysenko from deJong-Lambert, *Cold War Politics of Genetic Research*, 1–2.
70. On vernalization and early reactions to Lysenko, see Joravsky, *Lysenko Affair*; and Roll-Hansen, "New Perspective on Lysenko?," 261–78.
71. Lysenko, *Agrobiology*, 518.
72. Pringle, *Murder of Nikolai Vavilov*.
73. Carlson, *Genes, Radiation, and Society*, 223.
74. Rossianov, "Editing Nature," 728–45.
75. deJong-Lambert, *Cold War Politics of Genetic Research*, 52.
76. Sapp, *Beyond the Gene*, 171–72.
77. Weiner, "Roots of 'Michurinism,'" 243–60.
78. Graham, *Lysenko's Ghost*, 28.
79. Graham, 46.
80. Goncharov and Savel'ev, "Ivan V. Michurin," 105–27.
81. Graham, *Lysenko's Ghost*, 141.
82. Krementsov, "'Second Front' in Soviet Genetics," 229–50; and Harman, "C. D. Darlington and the British and American Reaction," 309–52.
83. Gordin, "How Lysenkoism Became Pseudoscience," 443–68.
84. Muller, "Artificial Transmutation of the Gene," 84–87.
85. Campos, "Dialectics Denied," 162.
86. Lysenko, *Agrobiology*, 548.
87. Campos, "Dialectics Denied," 178.
88. Saito, "Why Did Japanese Geneticists," 144–45.
89. Kihara and Sax, "Genetics in the U.S.S.R.," 158.
90. Iida, "Controversial Idea as a Cultural Resource," 562. Pressure to denounce Lysenko grew after 1950, when the Japanese Communist Party was outlawed.
91. Wolfe, "Cold War Context of the Golden Jubilee," 392.
92. Wolfe, 403.
93. Graham, *What Have We Learned*, 19.
94. Rushton, "William Bateson and the Chromosome Theory," 147–71.
95. von Schwerin, "Seeing, Breeding and the Organisation of Variation," 263.
96. Lysenkoists, for instance, sometimes referred to such organisms as "vegetative hybrids." *Graft hybrid* was also used as a catchall term at various points throughout the twentieth century for almost any grafted plant.
97. This focus has necessarily led to some omissions. For example, Lucien Louis Daniel, professor of

agricultural botany at the Lycée de Rennes in France, appears at multiple points in this book without a dedicated chapter. Daniel was seemingly disinterested in broader debates over graft hybrids and heredity, instead preferring to spend his time producing new plant varieties for horticulturalists.

98. Tilney-Bassett, *Plant Chimeras*, 7–8.

99. Foster and Aranzana, "Attention Sports Fans!," 3.

100. Hyun, *Bioethics and the Future of Stem Cell Research*, 108–9.

101. Tanaka, "Bizzarria—A Clear Case of Periclinal Chimera," 80–82.

Chapter 1: A Poultry Affair

1. Solberg, *Reforming Medical Education*, 163.

2. Guthrie, "Further Results of Transplantation," 565.

3. Hamilton, *First Transplant Surgeon*, 550–51.

4. Landecker, *Culturing Life*.

5. Friedman, *Immortalists*.

6. Cock and Forsdyke, *Treasure Your Exceptions*, 23–24.

7. William Bateson to family, 27 July (8 August) 1887, from Bateson and Bateson, *Letters from the Steppe*, 194. The two dates given correspond to the Julian and Gregorian calendars, respectively.

8. William Bateson to family, 23 August (4 September) 1887, from Bateson and Bateson, *Letters from the Steppe*, 203.

9. Cock and Forsdyke, *Treasure Your Exceptions*, 47.

10. Gregory Bateson, private communication to Arthur Koestler, 8 April 1970, from Koestler, *Case of the Midwife Toad*, 51.

11. Paul and Kimmelman, "Mendel in America," 283.

12. Bateson, "Hybridisation and Cross-Breeding," 59.

13. Cock and Forsdyke, *Treasure Your Exceptions*, 203.

14. Bateson, "Practical Aspects of the New Discoveries," 2.

15. Bateson, *Problems of Genetics*, 188.

16. Yandell Henderson to William Bateson, 1913, MS Add.8634/E.7, William Bateson, Scientific Correspondence and Papers, Special Collections, Cambridge University Library.

17. "Problems of Genetics," 497.

18. Bateson, *Problems of Genetics*, 189.

19. Paul and Kimmelman, "Mendel in America," 283.

20. For more on this argument, see Fitzgerald, *Business of Breeding*; and Thurtle, *Emergence of Genetic Rationality*.

21. On the founding and activities of the American Breeders' Association, see Kimmelman, "American Breeders' Association," 163–204.

22. Spillman, "Mendel's Law in Relation to Animal Breeding," 172.

23. Kimmelman, "American Breeders' Association," 172.

24. Scofield, "Description Forms and Score Cards," 27.

25. Buffum, "Effect of Environment on Plant Breeding," 216–19.

26. Burbank, "Heredity," 160.

27. Hansen, "Methods in Breeding Hardy Fruits," 168.

28. Hansen, 169.

29. Carleton, "Fundamental Requirements for Grain Breeding," 134.

30. Carleton, 134.

31. Buffum, "Effect of Environment on Plant Breeding," 212.

32. Simpson and Simpson, "Genetic Laws Applied," 250.

33. Simpson and Simpson, 255.

34. Simpson and Simpson, 254.

35. Guthrie, "Further Results of Transplantation," 567–69.

36. Guthrie, 571.

37. Dunn, "William Ernest Castle," 36.

38. Castle, "Mendel's Law of Heredity," 535.

39. Castle and Phillips, "Successful Ovarian Transplantation in the Guinea-Pig," 312.

40. Guthrie, "Guinea Pig Graft-Hybrids," 724.

41. Guthrie, 725.

42. Guthrie, "On Graft Hybrids," 358. The published version of Guthrie's address did not appear in the association's journal for another two years. Yet its title remained the same, suggesting that its content was not significantly altered.

43. Heape, "Preliminary Note on the Transplantation and Growth," 458.

44. Guthrie, "On Graft Hybrids," 358.

45. Guthrie, 369.

46. Castle and Phillips, *On Germinal Transplantation in Vertebrates*, 1.

47. Guthrie, "On Evidence of Soma Influence on Offspring," 816.

48. Guthrie, 816.

49. Castle, "On 'Soma Influence' in Ovarian Transplantation," 113.

50. Castle, 114.

51. Guthrie, "Transplantation of Ovaries," 918.

52. Charles Davenport to Charles Guthrie, 6 July 1907, Mss.B.D27, Charles Benedict Davenport Papers, Library of the American Philosophical Society, Philadelphia.

53. Witkowski, "Charles Benedict Davenport," 41.

54. MacDowell, "Charles Benedict Davenport," 26–28.

55. Davenport, "Mendel's Law of Dichotomy in Hybrids," 307.

56. Davenport, "Zoology of the Twentieth Century," 319.

57. Interpretation based on MacDowell, "Charles Benedict Davenport," 28; and Cock and Forsdyke, *Treasure Your Exceptions*, 295.

58. Curry, *Evolution Made to Order*, 18.

59. Charles Davenport to Charles Guthrie, 16 January 1908, Mss.B.D27, Correspondence, Davenport Papers.

60. Guthrie to Davenport, 20 January 1908, Mss.B.D27, Davenport Papers.

61. Davenport to Guthrie, 9 March 1908, Mss.B.D27, Davenport Papers.

62. Guthrie to Davenport, 15 April 1908, Mss.B.D27, Davenport Papers.

63. Davenport to Guthrie, 13 May 1908, Mss.B.D27, Davenport Papers.

64. Guthrie to Davenport, 19 May 1908, Mss.B.D27, Davenport Papers.

65. Davenport to Guthrie, 28 May 1908, Mss.B.D27, Davenport Papers.

66. Guthrie to Davenport, 22 August 1908, Mss.B.D27, Davenport Papers.

67. Davenport to Guthrie, 31 August 1908, Mss.B.D27, Davenport Papers.

68. Guthrie to Davenport, 14 April 1909, Mss.B.D27, Davenport Papers.

69. Davenport to Guthrie, 17 April 1909, Mss.B.D27, Davenport Papers.

70. Guthrie to Davenport, 29 April 1909, Mss.B.D27, Davenport Papers.

71. Guthrie, "Guinea Pig Graft-Hybrids," 724. Mss.B.D27, Davenport Papers.

72. Davenport, "Transplantation of Ovaries in Chickens," 122.

73. Guthrie to Davenport, 20 June 1911, Mss.B.D27, Davenport Papers.

74. Davenport, *Inheritance of Characteristics in Domestic Fowl*, 27.

75. Guthrie, "On Graft Hybrids," 369–71.

76. Huxley, *Soviet Genetics and World Science*, 6–7.

77. On the importance of model organisms to early genetics, see Müller-Wille and Rheinberger, *Cultural History of Heredity*, 127–28.

78. Collins, "Son of Seven Sexes," 34.

79. Collins, 34.

80. Crew, *Animal Genetics*, 339.

Chapter 2: Rise of the Chimera

1. Charles Davenport to George Harrison Shull, 1 November 1910, folder 3, 1909–1911, Mss.B.D27, Correspondence, Charles Benedict Davenport Papers, Library of the American Philosophical Society, Philadelphia.

2. Shull to Davenport, 8 November 1910, Mss.B.D27, Davenport Papers.

3. Harwood, *Styles of Scientific Thought*, 35.

4. Cowles and Chamberlain, "Graft Hybrids and Chimeras," 147–53.

5. von Schwerin, "Seeing, Breeding and the Organisation of Variation," 261–63.

6. On Bailey's early recognition of Mendel, see Zirkle, "Role of Liberty Hyde Bailey," 205–18.

7. Bailey, *Cyclopedia of American Horticulture*, 661.

8. Smith, "Douglas Houghton Campbell," 46–47.

9. Gates, "Graft Hybrids," *Botanical Gazette* 49, no. 5 (1910), 387.

10. Campbell, "Nature of Graft-Hybrids," 45.

11. Campbell, 48–49.

12. Campbell, 52.

13. Campbell, 51.

14. Campbell, 52.

15. Campbell, *Plant Life and Evolution*, 312–13.

16. Campbell, 317–18.

17. Campbell, 329.

18. Roberts, "Reginald Ruggles Gates," 84–87.

19. Gates, "Graft Hybrids," *Botanical Gazette* 47, no. 1 (1909), 84.

20. Gates, 84.

21. Gates, "Graft Hybrids," *Botanical Gazette* 47, no. 3 (1909), 250.

22. Gates, "Mendelism," 61–62.

23. Gates, "Graft Hybrids," *Botanical Gazette* 48, no. 6 (1909), 478.

24. Gates, "Graft Hybrids," *Botanical Gazette* 49, no. 5 (1910), 387.

25. William Bateson to R. R. Gates, 19 May 1911, MS Add.8634/H.13, William Bateson Papers, Special Collections, Cambridge University Library.

26. Bateson to Gates, 2 March 1911, MS Add.8634/H.13, Bateson Papers.

27. Castle, "Apple Chimera," 202.

28. Richmond, "Women in the Early History of Genetics," 59–63.

29. Richmond, 84.

30. Cock and Forsdyke, *Treasure Your Exceptions*, 383.

31. William Bateson to Beatrice Bateson, 06 January 1909, MS Add.8634/A.38:3.1, Bateson Papers.

32. William Bateson to Erwin Baur, 30 December 1909, MS Add.8634/H.1, Bateson Papers.

33. Bateson to Baur, 8 May 1910, MS Add.8634/H.1, Bateson Papers.

34. Bateson to Baur, 31 March 1911, MS Add.8634/H.1, Bateson Papers.

35. Bateson to Baur, 13 April 1911, MS Add.8634/H.1, Bateson Papers.

36. Bateson to Baur, 13 April 1911, MS Add.8634/H.1, Bateson Papers.

37. Bateson to Baur, 29 April 1911, MS Add.8634/H.1, Bateson Papers.

38. Bateson to Baur, 13 May 1911, MS Add.8634/H.1, Bateson Papers.

39. Cock and Forsdyke, *Treasure Your Exceptions*, 407.

40. Bateson to Baur, 31 January 1913, MS Add.8634/H.1, Bateson Papers.

41. Bateson to Baur, 25 February 1913, MS Add.8634/H.1, Bateson Papers.

42. Bateson to Baur, 9 May 1913, MS Add.8634/H.1, Bateson Papers.

43. William Bateson to Signe Laura Amalia Nilsson-Ehle, 7 April 1916, MS Add.8634/H.29, Bateson Papers.

44. Bateson, "Root-Cuttings, Chimaeras and 'Sports,'" 75.

45. Bateson, 75–76.

46. Bateson, 76.

47. Harvey, "Pioneers of Genetics," 105.

48. Radick, *Disputed Inheritance*, 291.

49. Harvey, "Pioneers of Genetics," 115.

50. Cock and Forsdyke, *Treasure Your Exceptions*, 593.

51. Olby, "Bateson, William," 505–6; and Rushton, "William Bateson and the Chromosome Theory," 147–71.

52. Forrester and Cameron, *Freud in Cambridge*, 48–49.

53. Boulter, *Bloomsbury Scientists*, 78–80.

54. Porter, "Vernon Herbert Blackman," 39.

55. Blackman, "Some Recent Work on Hybrids in Plants," 80.

56. Letter to Arthur Tansley, 21 August 1912, MS Tansley H.7, Arthur Tansley Papers, Special Collections, Cambridge University Library.

57. R.P.G., "Graft-Hybrids," 214.

58. Hume, "On the Presence of Connecting Threads," 217.

59. Macfarlane, "Comparison of the Minute Structure," 270.

60. Macfarlane, "Observations on Some Hybrids," 241–49.

61. Hume, "On the Presence of Connecting Threads," 217.

62. Hume, 219.

63. Hume, 220.

64. Popenoe, "Plant Chimeras," 521.

65. Popenoe, 532.

66. Swingle, "Graft Hybrids in Plants," 86.

67. Smith, "Chromosome Counts in the Varieties," 84–92.

68. Harwood, *Styles of Scientific Thought*, 65–66.

69. Popenoe, "Plant Chimeras," 532.

70. Shull, "Graft-Hybrids," 359.

71. Shull, "'Graft Hybrids' of Bronvaux," 324.

72. Harwood, *Styles of Scientific Thought*, 49–50.

73. Skene, "Plant Chimæras," 127–29.

74. Skene, 132–34.

Chapter 3: Creed in the Place of Science

1. "Graft-Inheritance," 174.

2. Johannsen, "Inheritance of Characters Acquired by Grafting," 536.

3. On the scientific history of the chromosome, see Chadarevian, *Heredity under the Microscope*.

4. Howard, review of *Plant Chimeras*, 503.

5. "Scientist Tells of Success Where Darwin Met Failure," *New York Times*, 3 June 1923, MS Add.8634/C.30:6.1v, Scientific Topics and Controversies, William Bateson Papers, Special Collections, Cambridge University Library.

6. William Bateson to Beatrice Bateson, 28 September 1910, MS Add.8634/C.28:3.2, Bateson Papers.

7. Logan, *Hormones, Heredity, and Race*, 43.

8. Coen, "Living Precisely in Fin-de-Siècle Vienna," 497–98.

9. Gliboff, "Case of Paul Kammerer," 533–34.

10. Kammerer, *Inheritance of Acquired Characteristics*, 125.

11. Gliboff, "Case of Paul Kammerer," 547–48.

12. Gliboff, 548.

13. MacBride, "Variety and Environment in Lizards," 71.

14. Gliboff, "Case of Paul Kammerer," 551.

15. Kammerer, "Breeding Experiments," 637.

16. Kammerer, *Inheritance of Acquired Characteristics*, 96–101.

17. Kammerer, 102–5.

18. Kammerer, 138–39.

19. Kammerer, "Breeding Experiments," 638.

20. Kammerer, *Inheritance of Acquired Characteristics*, 144.

21. Kammerer, 147.

22. Bateson, *Problems of Genetics*, 209.

23. Bateson, "Dr. Kammerer's Testimony," 344.

24. Koestler, *Case of the Midwife Toad*, 68.

25. Gliboff, "Case of Paul Kammerer," 542.

26. For more on Cunningham, see Bowler, *Eclipse of Darwinism*.

27. Cunningham, "Breeding Experiments," 702.

28. Kammerer, *Inheritance of Acquired Characteristics*, 287.

29. Kammerer, 288–91.

30. Koestler, *Case of the Midwife Toad*, 114–15. Klaus Taschwer suggests that Kammerer may have fallen victim to an antisemitic group of professors at Vienna University, who operated under the codename "bear's lair." See Taschwer, *Der Fall Paul Kammerer*.

31. For example, see Whittaker, "Siphon Regeneration in *Ciona*," 224–25.

32. Weiss, "Life," 65.

33. Weiss, 66.

34. Thomas, "Frederick Ernest Weiss," 601–2.

35. Thomas, 603–4.

36. Weiss, "Researches on Heredity in Plants," 3.

37. Weiss, 11. This apparent success story for Mendelian genetics was not quite as straightforward as Weiss portrayed it. The Swedish plant breeder Herman Nilsson-Ehle would later confirm that immunity in wheat was not a simple Mendelian character and could fluctuate unpredictably from generation to generation.

38. Weiss, 12.

39. Alberti, "Amateurs and Professionals in One County," 120.

40. Weiss, "Microscopy in Manchester," 43–44.

41. Weiss, 46.

42. Weiss, 44–46.

43. Weiss, "Graft Hybrids," 32.

44. Weiss, 34.

45. Weiss, 37.

46. Weiss, 39.

47. Weiss, 39–40.

48. Onaga, "Toyama Kametaro and Vernon Kellogg," 243–44.

49. Weiss, "Graft Hybrids," 40.

50. L. A. Boodle to F. E. Weiss, 25 September 1916, MS/319/6, Correspondence of Frederick Ernest Weiss, Linnean Society Archives, London.

51. N. Wille to F. E. Weiss, 10 October 1916, MS/319/52, Weiss Correspondence. A later letter implies that Weiss considered visiting Norway, perhaps to view the graft hybrid. It is unknown whether this trip went ahead. See Wille to Weiss, 31 October 1916, MS/319/53, Weiss Correspondence.

52. Weiss, "On the Leaf-Tissues of the Graft Hybrids," 78.

53. Weiss, "Some Recent Advances in Our Knowledge," 75.

54. Chittenden, "Vegetative Segregation," 413.

55. Chittenden, 425.

56. Weiss, "Problem of Graft Hybrids and Chimaeras," 265–66.

57. Weiss, 264.

58. Weiss, 264.

59. Weiss, "Graft Hybrids and Chimaeras: I.," 215.

60. Weiss, 217.

61. Weiss, "Graft Hybrids and Chimaeras: II.," 243.

62. Creese and Creese, "British Women Who Contributed to Research," 36.

63. Haines, *International Women in Science*, 258.

64. Neilson Jones and Rayner, *Textbook of Plant Biology*, v.

65. Neilson Jones and Rayner, 170.

66. Neilson Jones and Rayner, 171.

67. Neilson Jones and Rayner, 183.

68. Neilson Jones and Rayner, 184.

69. Neilson Jones, *Plant Chimaeras and Graft Hybrids*, 3.

70. Neilson Jones, 6–7.

71. Neilson Jones, 29–30.

72. Neilson Jones, 30.

73. Neilson Jones, 31.

74. Neilson Jones, 33–34.

75. Neilson Jones, 44.

76. Neilson Jones, 45–46.

77. Neilson Jones, 56.

78. Salisbury, review of *Plant Chimæras and Graft Hybrids*, 173.

79. Small, review of *Plant Chimaeras and Graft Hybrids*, 94.

80. Weiss, review of *Plant Chimaeras and Graft Hybrids*, 391.

81. Neilson Jones, "Chimaeras," 558.

82. Neilson Jones, 560.

83. Neilson Jones, 561.

84. Neilson Jones, *Growing Plant*, 108.

85. Howard, review of *Plant Chimeras*, 503.

86. Clowes, review of *Plant Chimeras*, 1252.

87. Neilson Jones, *Plant Chimaeras and Graft Hybrids*, 31.

88. Neilson Jones, 12.

89. This was not the case for transplantation more broadly. In embryology and genetics, transplanting tissue had become an established technique in laboratory experiments. See Brandt, "Development and Heredity in the Interwar Period," 253–83.

90. Dierig, Lachmund, and Mendelsohn, "Toward an Urban History of Science," 2.

91. Coen, "Living Precisely in Fin-de-Siècle Vienna," 497.

92. Coen, 512–13.

93. Gliboff, "Case of Paul Kammerer," 551.

94. Stráner, "Natural Sciences and Their Public," 59–79.

Chapter 4: Beyond Lysenko

1. Gould, *Hen's Teeth and Horse's Toes*, 135.

2. Lysenko, *Agrobiology*, 545–46.

3. Cook, "Lysenko's Marxist Genetics," 184. Lysenko's use of wax models led some in the Western press to accuse him of fraud. However, Lysenko was—in this instance—defended by British biologist Julian Huxley, who argued that the use of wax models for botanical displays was normal practice.

4. A complete list of the literature on the Lysenko affair would be large enough to fill an entire volume. Two classic introductions to the affair are Graham, *Science and Philosophy in the Soviet Union* and Krementsov, *Stalinist Science*.

5. Roll-Hansen, *Lysenko Effect*.

6. See Goncharov and Savel'ev, "Ivan V. Michurin," 105–27; and Weiner, "Roots of 'Michurinism,'" 243–60.

7. Wolfe, "What Does It Mean to Go Public?," 48–78.

8. Roll-Hansen, "Lamarckism and Lysenkoism Revisited," 85–86.

9. Cook, "Lysenko's Marxist Genetics," 184.

10. Werskey, *Visible College*; and Werskey, "Visible College Revisited," 305–19.

11. For a more sympathetic perspective on Haldane's lackluster defense of Lysenko and the BBC debate, see deJong-Lambert, "H. J. Muller and J. B. S. Haldane," 103–35.

12. Palló and Müller, "Opportunism and Enforcement," 3–36.

13. Subramanian, *Dominant Character*, 63.

14. J. B. S. Haldane, "Why I Am Cooperator," unpublished manuscript, 1942, f. 10, Haldane Box 4, HALDANE/1/2/63, J. B. S. Haldane Papers, University College London (UCL) Archives, accessed through UCL Digital Collections, https://www.ucl.ac.uk/library/digital-collections.

15. Subramanian, *Dominant Character*, 63.

16. Haldane, "Why I Am Cooperator."

17. Clark, *J.B.S.*, 66–68.

18. See Harman, *Man Who Invented the Chromosome*.

19. Wilmot, "J. B. S. Haldane," 819.

20. Subramanian, *Dominant Character*, 166.

21. Subramanian, 41.

22. Clark, *J.B.S.*, 115.

23. Haldane, "Why I Am Cooperator."

24. Subramanian, *Dominant Character*, 223.

25. Russian embassy and J. B. S. Haldane, correspondence, December 1944–January 1945, HALDANE/4/9/1/1, Haldane Box 27, Haldane Papers.

26. Haldane, *Science Advances*, 224.

27. Haldane, 224.

28. Haldane, 225.

29. J. B. S. Haldane, "Plants and Animals in Human History," 1936, HALDANE/1/2/172, f. 1. Haldane Box 5, Haldane Papers.

30. J. B. S. Haldane, "Synthetic Food" (n.d., mid-twentieth century), HALDANE/1/2/103, f. 3. Haldane Box 4a, Haldane Papers.

31. Haldane, "Synthetic Food."

32. J. B. S. Haldane, essay on food production methods (n.d., mid-twentieth century), HALDANE/1/2/100, f. 1. Haldane Box 4a, Haldane Papers.

33. Haldane, essay on food production methods. For more on Haldane's views on synthetic food and his frustration with science policy in Britain, see Holmes, "Yeast, Coal, and Straw," 202–20.

34. Bengtsson and Tunlid, "1948 International Congress of Genetics in Sweden," 709–15.

35. J. B. S. Haldane, article on plant and animal genetics, 1948, HALDANE/2/1/2/98, f. 2. Haldane Box 9, Haldane Papers.

36. Haldane, article on plant and animal genetics.

37. Subramanian, *Dominant Character*, 169.

38. Cornforth, *Communism and Philosophy*, 7.

39. A. G. Morton, draft statement, "The Present Position in Biology," 19 November 1948, HALDANE/4/9/1/4, f. 4. Haldane Box 27, Haldane Papers.

40. J. B. S. Haldane and Maurice Cornforth, correspondence, November 1948–December 1948, HALDANE/4/9/1/5, f. 1. Haldane Box 27, Haldane Papers.

41. "In Support of Lysenko," discussion statement, December 1948, HALDANE/4/9/1/7, f. 1. Haldane Box 27, Haldane Papers.

42. "In Support of Lysenko."

43. "In Support of Lysenko."

44. Haldane, "Some Alternatives to Sex," 22.

45. Haldane, 23.

46. Haldane, 23.

47. Paul, "War on Two Fronts," 1–37.

48. Dewsbury, "Darwin-Bateman Paradigm in Historical Context," 831–37.

49. A. J. Bateman to J. B. S. Haldane, 1938–1939, HALDANE/5/2/1/40, f. 1. Haldane Box 42, Haldane Papers.

50. Bateman to Haldane, 1938–1939, HALDANE/5/2/1/40, f. 1–2. Haldane Box 42, Haldane Papers.

51. Bateman to Haldane, 1938–1939, HALDANE/5/2/1/40, f. 3. Haldane Box 42, Haldane Papers.

52. Bateman to Haldane, 1938–1939, HALDANE/5/2/1/40, f. 5. Haldane Box 42, Haldane Papers.

53. Bateman to Haldane, 1938–1939, HALDANE/5/2/1/40, f. 11–12. Haldane Box 42, Haldane Papers.

54. On Darlington, see Harman, *Man Who Invented the Chromosome*.

55. Bateman to Haldane, 1938–1939, HALDANE/5/2/1/40, f. 6. Haldane Box 42, Haldane Papers.

56. Bateman to Haldane, 29 August 1948, HALDANE/5/1/2/8/17, f. 1r. Haldane Box 38, Haldane Papers.

57. Bateman to Haldane, 29 August 1948, HALDANE/5/1/2/8/17, f. 1v.

58. Haldane to Bateman, 23 September 1948, HALDANE/5/1/2/8/23, f. 1-2. Haldane Box 38, Haldane Papers.

59. A. J. Bateman, discussion statement on the Lysenko controversy, 7 December 1948, HALDANE/4/9/1/6, f. 1. Haldane Box 27, Haldane Papers.

60. Bateman, discussion statement on the Lysenko controversy.

61. A. J. Bateman, "The Genetics Controversy in the Soviet Union," discussion statement, January 1949, HALDANE/4/9/1/8, f. 1. Haldane Box 27, Haldane Papers.

62. Bateman, "Genetics Controversy in the Soviet Union."

63. Bateman, "Genetics Controversy in the Soviet Union."

64. Bateman to Haldane, 9 February 1949, HALDANE/4/9/2/10, f. 1. Haldane Box 27, Haldane Papers.

65. McLaren and Michie, "Current Trends of Genetical Research in Hungary," 390–91.

66. Casselton and Jones, "Dan Lewis," 171.

67. Bateman, "Grafting Experiments between the Tomato Varieties," 1118.

68. Bateman, 1119.

69. Bateman, 1119.

70. Bateman, 1119.

71. Felföldy, "Tomato Grafting Experiments at Tihany," 1144.

72. Felföldy, 1144.

73. Darlington, review of *Genetics of the Dog*, 141.

74. Burns, *Genetics of the Dog*, 85.

75. Burns, 86.

76. Darlington, review of *Genetics of the Dog*, 142.

77. Darlington, 141.

78. Hudson and Richens, *New Genetics in the Soviet Union*, 45.

79. Hudson and Richens, 46.

80. Hudson and Richens, 46.

81. Hudson and Richens, 48.

82. Marca Burns to J. B. S. Haldane, December 1940–February 1941, HALDANE/5/2/1/98, Haldane Box 42, Haldane Papers.

83. British Cattle Breeders' Club (Monck to Haldane), April 1948, HALDANE/4/26/8, f. 4. Haldane Box 36, Haldane Papers.

84. British Cattle Breeders' Club (Monck to Haldane).

85. On the history of the Royal Horticultural Society, see Fletcher, *Story of the Royal Horticultural Society*.

86. Crane, review of *Selected Works*, 369. Writing on the inheritance of acquired characters in 1934, Michurin wrote that for "us fruit growers our entire case [in favor of acquired characters] is usually based on the propagation of new hybrid varieties by the vegetative method of grafting or cutting." Yet he also acknowledged that the heritable changes induced by grafting were slight. See Michurin, *Some Problems of Method*, 54–55.

87. Crane, review of *Selected Works*, 370.

88. Sirks, "Royal Horticultural Society," 216.

89. Janaki Ammal, "Chromosomes and Horticulture," 236.

90. Janaki Ammal, 239.

91. Knight, *Abstract Bibliography of Fruit Breeding*, foreword.

92. Knight, 199.

93. Knight, 251.

94. Knight, 408–9.

95. Tompsett, "Fruit Grower Visits the U.S.S.R.," 354.

96. Tompsett, 355.

97. Tompsett, 356.

98. Tompsett, 358.

99. Tompsett, 358–59.

100. Tompsett, 359.

101. Tompsett, 361.

102. Hughes, "Modern Techniques in Fruit Growing," 223.

103. Jinks, *Extrachromosomal Inheritance*, 1.

104. Jinks, 21–24.

105. Jinks, 72.

106. "Heredity Mechanisms and Evolution," *Times* (London), 23 April 1965.

107. J. H. Oliver, "Proctor," reprinted from the *Brewers' Guild Journal*, April 1957, file 1, box 26, Plant Breeding Institute, G. B. H. Bell Correspondence, John Innes Centre, Norwich, England. Beaven was part of an early twentieth-century backlash against Mendelian genetics in Britain. See Palladino, *Plants, Patients and the Historian*, 79–81 and Kingsbury, *Hybrid*, 173–74.

108. Holmes, "Crops in a Machine," 146.

109. Kolchinsky, "Current Attempts at Exonerating 'Lysenkoism,'" 224.

Chapter 5: Transplantation and Tomatoes

1. Donald Michie, reprint of article "The Moscow Institute of Genetics" from *Discovery* magazine (1957), 432. Michie Misc. Articles, 83980, Anne McLaren Papers, British Library, London.

2. Hogan, "From Embryo to Ethics," 477–82.

3. Anne McLaren, "Social Context of the New Genetics," 1, 83939, McLaren Papers. The text indicates that this talk was written a few months after a Royal Society Discussion Group on genetic modification in July 1992.

4. McLaren, "Social Context of the New Genetics," 2.

5. Medawar, *Memoir of a Thinking Radish*, 77.

6. Hamilton, *History of Organ Transplantation*, 223.

7. Billingham, Brent, and Medawar, "'Actively Acquired Tolerance' of Foreign Cells," 603.

8. Brent, *History of Transplantation Immunology*, 201.

9. Paleček, "Vítězslav Orel," 4.

10. Ivanyi, "Milan Hašek and the Discovery of Immunological Tolerance," 591–92.

11. McLaren, "International Rapprochement, 50 Years Ago," 1425.

12. Ivanyi, "Milan Hašek and the Discovery of Immunological Tolerance," 594.

13. Illman, "Leslie Brent."

14. Brent, *History of Transplantation Immunology*, 201.

15. Hašek, "Tolerance Phenomena in Birds," 67.

16. Billingham, Brent, and Medawar, "Quantitative Studies on Tissue Transplantation Immunity," plates 7 and 8, facing 413–14.

17. Billingham, Brent, and Medawar, 359.

18. Billingham, Brent, and Medawar, 360.

19. McLaren and Michie, "Current Trends of Genetical Research in Hungary," 390–91.

20. Michie and McLaren, "Importance of Being Cross-Bred," 65.

21. McLaren, "International Rapprochement, 50 Years Ago," 1425.

22. Anne McLaren to Juraj Ivanyi, 16 December 2002, 89202/3/4, McLaren Papers.

23. Ivanyi, "Milan Hašek and the Discovery of Immunological Tolerance," 594. Hašek and McLaren continued their correspondence well into the 1960s. Milan Hašek to Anne McLaren, 22 November 1967, 89202/2/2, McLaren Papers.

24. McLaren, "Social Context of the New Genetics," 1, 83939, McLaren Papers.

25. Herbert Butler Parry to Anne McLaren, 4 August 1960, 89202/4/1, McLaren Papers.

26. Medawar, "Is the Scientific Paper a Fraud?," 377–78.

27. Calver, "Sir Peter Medawar," 301–14.

28. Anne McLaren to Professor Harland, 1 January 1958, 89202/4/1, McLaren Papers.

29. Duančić, "Lysenko in Yugoslavia, 1945–1950s," 172–74.

30. Duančić, 185.

31. Victor [ARC] to Anne McLaren, 21 February 1958, 89202/4/1, McLaren Papers.

32. Anne McLaren, A Proposed Experiment on Graft Hybridisation in Tomatoes, 1, 89202/2/5, McLaren Papers.

33. McLaren, Proposed Experiment on Graft Hybridisation in Tomatoes, 1.

34. McLaren, Proposed Experiment on Graft Hybridisation in Tomatoes, 2.

35. Groner, Sachs, and Lotem, "Leo Sachs," 355–75.

36. Sachs, "Vegetative Hybridization," 1010.

37. McLaren, Proposed Experiment on Graft Hybridisation in Tomatoes, 2.

38. McLaren to Harland, 1 January 1958, 89202/4/1, McLaren Papers.

39. McLaren, Proposed Experiment on Graft Hybridisation in Tomatoes, 2–4.

40. D. Lewis to Anne McLaren, 21 March 1958, 89202/4/1, McLaren Papers.

41. Lewis to McLaren, 2 February 1959, 89202/4/1, McLaren Papers.

42. Ružicka Glavinic to Anne McLaren, 12 January 1959, 89202/2/5, McLaren Papers.

43. Anne McLaren, Tomato Experiment, Visit to Belgrade, May 1958, 89202/2/5, McLaren Papers.

44. Schneider, "Michurinist Biology in the People's Republic," 526.

45. Schneider, 527.

46. Tsu and Chao, "Study on the Vegetative Hybridization of Some Solanaceous Plants," 889.

47. Tsu and Chao, 889–90.

48. Tsu and Chao, 891.

49. Tsu and Chao, 892–93.

50. Tsu and Chao, 899.

51. Tsu and Chao, 901–2.

52. Tsu and Chao, 903.

53. Schneider, *Biology and Revolution in Twentieth-Century China*, 156.

54. Schneider, 173.

55. Schneider, 205.

56. Prof. T. M. Tsu, Chinese Academy of Agricultural Science, Peking West, to Anne McLaren, 1 April 1958, 89202/3/2 5075A, McLaren Papers.

57. Michie and McLaren, "Importance of Being Cross-Bred," 65.

58. Gordin, "Lysenko Unemployed," 56–78.

59. Donald Michie, "Interview with Lysenko," *SCR Soviet Science Bulletin* 5 (1958): 5, 83980, McLaren Papers.

60. Michie, "Interview with Lysenko," 6.

61. Franklin, "Obituary," 856.

62. Crew, *Animal Genetics*, 339.

63. These experiments are mentioned in Waddington's early laboratory notes (probably dating to his time in Cambridge). See, for example: Laboratory notebook relating mainly to chick embryological and grafting experiments, ca. 1930–1946; laboratory notebooks and bundles of research notes, ca. 1930–1974, Coll-41/1/1, papers of Conrad Hal Waddington, Special Collections, Edinburgh University Library.

64. Conrad Hal Waddington to J. Needham, 20 February 1965, personal correspondence, vol. 2, 1962–1965, Coll-41/9/4/2, Waddington Papers.

65. Franklin, "Obituary," 858.

66. Anne McLaren, report on visit to Beijing, 23 September–2 October 1985, 83939, McLaren Papers.

67. McLaren, "Social Context of the New Genetics," 3, 83939, McLaren Papers.

68. McLaren, "Too Late for the Midwife Toad," 169.

69. Anne McLaren, "Mendel and Michurin Today," 2005, 3, 89202/2/6, McLaren: Material concerning McLaren's work on Mendel and Michurin, McLaren Papers.

70. Ohta, "Variant Found in the Progeny from Grafting," 34–35.

71. Ohta and Chuong, "Hereditary Changes in *Capsicum Annuum* L.," 356.

72. Ohta and Chuong, 366.

73. Ohta and Chuong, 368.

74. Ohta, "Tsugiki de kawaru iden keishitsu," 100–113.

75. Ohta, "Graft-Transformation, the Mechanism for Graft-Induced Genetic Changes," 91–99.

76. McLaren, "Mendel and Michurin Today," 3, 89202/2/6, McLaren: Material concerning McLaren's work on Mendel and Michurin, McLaren Papers.

77. McLaren, "Mendel and Michurin Today," 7.

78. Anne McLaren to Robert Shields, 4 March 2005, 89202/4/38, McLaren Papers.

79. Shields to McLaren, 3 March 2005, 89202/4/38, McLaren Papers.

80. Yongsheng Liu, email to Anne McLaren, 28 September 2003, 89202/3/3, McLaren Papers.

81. Liu, "Historical and Modern Genetics," 103.

82. Liu, 105–7.

83. Roll-Hansen, "Lamarckism and Lysenkoism Revisited," 85.

84. Liu, "Historical and Modern Genetics," 118.

85. Liu, 120.

86. Michurin, *Selected Works*.

87. Liu, "Historical and Modern Genetics," 122.

88. Liu, 125.

89. This account was written by Yongsheng Liu on an online obituary notice for Anne McLaren: "Anne McLaren DBE, DPhil, FRS, FRCOG: April 26th 1927–July 7th 2007," Gurdon Institute, last updated 5 August 2015, http://www2.gurdon.cam.ac.uk/anne-mclaren.html.

90. Liu, "Expanding the Potential of Plant Interfamily Grafting," 448–49.

91. McLaren, "Too Late for the Midwife Toad," 169.

Chapter 6: Cell Fusion

1. Agar, "What Happened in the Sixties?," 570.

2. Galston, "Molecular Biology and Agricultural Botany," 158.

3. Galston, 159.

4. References to somatic hybridization occur in Kloppenburg, *First the Seed*, 192; Charles, *Lords of the Harvest*, 9; Lurquin, *Green Phoenix*, 102; Schurman, "Biotechnology in the New Millennium," 20; and Holmes, "Somatic Hybridization," 1–23.

5. Munns, "Phytotronist and the Phenotype," 29. On the contribution of plant physiologists to molecular biology, see Zallen, "Redrawing the Boundaries of Molecular Biology," 65–87.

6. Kenney, *Biotechnology*; Hughes, "Making Dollars Out of DNA," 541–75; Kleinman, *Impure Cultures*; and Rasmussen, *Gene Jockeys*.

7. Munns, "Phytotronist and the Phenotype," 29.

8. Cocking, "Plant Protoplasts" (2000), 77.

9. Cocking, "Plant Protoplasts" (1965), 170–203. For the original nineteenth-century paper, see Klercker, "Method for the Isolation of Living Protoplasts," 463–74.

10. Protoplasts can occur naturally, allowing fusion between plant cells to occur. Küster observed "naked vacuolar membranes" in the sap of solanaceous berries. Küster, "Über die Gewinnung nackter Protoplasten," 223–33. Decades later, protoplasts, protoplasmic units, and vacuoles were observed in tomato fruit locale tissue. Cocking and Gregory, "Organized Protoplasmic Units of the Plant Cell," 504–11.

11. Smith, "Chromosome Counts in the Varieties," 84–92.

12. Plowe, "Membranes in the Plant Cell," 197–98. Plowe saw herself within a tradition of cell research and micromanipulation, beginning with de Vries's 1885 study of the tonoplast (the layer of cytoplasm around the plant vacuole). Plowe favored the term *micromanipulation* over *microdissection* for her work, as the latter implied the study of dead organisms.

13. Plowe, 196.

14. Another important development came in the form of tissue culture, required to "regenerate" fused cells in a viable plant. See Landecker, *Culturing Life*.

15. Küster, "Über die Gewinnung nackter Protoplasten," 223–34.

16. Michel, "Uber die experimentelle Fusion pflanzlicher Protoplasten," 230–52.

17. Neilson Jones, *Plant Chimaeras and Graft Hybrids*, 33.

18. Munns, "Phytotronist and the Phenotype," 32.

19. Landecker, *Culturing Life*, 188.

20. Wilson, *Tissue Culture in Science and Society*, 71.

21. Wilson, 71–72.

22. Okada, "Cell Fusion and Somatic Cell Genetics," 143.

23. Harris et al., "Mitosis in Hybrid Cells," 606–8.

24. Wilson, *Tissue Culture in Science and Society*, 70.

25. Harris, "Hybrid Cells from Mouse and Man," 358.

26. Boveri, "Concerning the Origin of Malignant Tumours," 52.

27. Landecker, *Culturing Life*, 199.

28. Landecker, 153. A similar trend can be seen with recombinant DNA technology, which was first applied to bacterial and animal cells before its use in plants.

29. Harris, *Cells of the Body*, 142.

30. On the media portrayal of cell fusion, see Wilson, *Tissue Culture in Science and Society*, 75.

31. John F. Watkins, "Scientists and Society," *Times* (London), 16 December 1967, 7.

32. Donald Fleming, "On Living in a Biological Revolution," *Atlantic Monthly*, February 1969, 64–65.

33. "Cell Fusion," 1039.

34. "Cell Fusion," 1039–40.

35. "Cell Fusion," 1041.

36. Klein and Klein, "How One Thing Has Led to Another," 9–10. The Soviet Academy refused to accept Muller's resignation and expelled him.

37. Klein and Klein, 28.

38. Cocking, "Properties of Isolated Plant Protoplasts," 780.

39. Cocking, 781.

40. Frey-Wyssling, "Gymnoplasts Instead of 'Protoplasts,'" 516.

41. For an English translation of Brabec's entry, see Constabel, "Somatic Hybridization in Higher Plants," 743.

42. Galston, "Molecular Biology and Agricultural Botany," 158.

43. Ehrlich's work formed part of a postwar Malthusian literature. Schoijet, "*Limits to Growth and the Rise of Catastrophism*," 515–30; and Robertson, *Malthusian Moment*.

44. Galston, "Molecular Biology and Agricultural Botany," 159.

45. Nickell and Torrey, "Crop Improvement," 1068.

46. Nickell and Torrey, 1070.

47. Nagata and Takebe, "Cell Wall Regeneration and Cell Division," 303–4.

48. Nagata and Takebe, 307.

49. Takebe, Labib, and Melchers, "Regeneration of Whole Plants," 320.

50. Cocking, "Properties of Isolated Plant Protoplasts," 782.

51. Cocking, "Plant Protoplasts" (2000), 78.

52. Carlson, Smith, and Dearing, "Parasexual Interspecific Plant Hybridization," 2292.

53. Carlson, Smith, and Dearing, 2293.

54. Arthur Bennett and Ted Schoeters, "Birds and Bees Not Wanted," *Financial Times*, 23 August 1972, 8.

55. Carlson, Smith, and Dearing, "Parasexual Interspecific Plant Hybridization," 2294.

56. Bennett and Schoeters, "Birds and Bees Not Wanted," 8.

57. Cocking et al., "Selection Procedures for the Production," 7–12.

58. Constabel, "Somatic Hybridization in Higher Plants," 747.

59. Wharton was a member of the Presidential Commission on World Hunger, which delivered its report to US president Jimmy Carter in March 1980. Wharton, "Food, the Hidden Crisis," 1415.

60. Bhojwani, Evans, and Cocking, "Protoplast Technology in Relation to Crop Plants," 351.

61. Bhojwani, Evans, and Cocking, 356.

62. Cocking, "Opportunities from the Use of Protoplasts," 557. Some of these varieties had been created not through somatic hybridization but through cytoplasmic hybridization, which can involve the movement of organelles or transfer of nuclei via a fusion event.

63. Cocking, 566.

64. Mather, "Perspective and Prospect," 607.

65. Bevan, Flavell, and Chilton, "Chimaeric Antibiotic Resistance Gene," 184–87.

66. Shepard et al., "Genetic Transfer in Plants," 683–88.

67. Fowler, "Plant Cell Biotechnology and Agriculture," 215.

68. Fowler, 220.

69. Li, "Advances in Plant Genetic Manipulation," 219–20.

70. Keller et al., "Application of Somatic Hybridization Technology," 192–93.

71. Simmonds and Smartt, *Principles of Crop Improvement*, 290.

72. Scowcroft, "Genetic Manipulation in Crops," 13.

73. Scowcroft, 15.

74. Simmonds and Smartt, *Principles of Crop Improvement*, 288.

75. Nagata and Takebe, "Cell Wall Regeneration and Cell Division," 303–4.

76. Cocking, "Plant Protoplasts" (2000), 78.

77. Williams, "Irene Manton, Erwin Schrödinger," 425–59.

78. Cocking, "Plant Protoplasts" (2000), 78–79.

79. Cocking, 78.

80. Keith Roberts to James D. Watson, 18 November 1974, JDW/2/2/1550/52, James D. Watson Collection, Archives Repository, Cold Spring Harbor Laboratory.

81. Carlson, Smith, and Dearing, "Parasexual Interspecific Plant Hybridization," 2292.

82. Carlson, Smith, and Dearing, 2292–93.

83. Carlson, Smith, and Dearing, 2293.

84. Carlson, Smith, and Dearing, 2292. On electrophoresis, see Holmes, "Changing Techniques in Crop Plant Classification," 149–64.

85. Carlson, Smith, and Dearing, "Parasexual Interspecific Plant Hybridization," 2293–94.

86. Gleba, Sytnik, and Shoeman, *Protoplast Fusion*, 188.

87. Gleba, Sytnik, and Shoeman, 188. Cocking reviewed Gleba and Sytnik's monograph and described it as "essential reading." Cocking referred to the author's call for an "instillation of genetic ideology" as "unfortunate phraseology." Cocking, review of *Protoplast Fusion*, 432.

88. Hughes, "Making Dollars Out of DNA," 542.

89. Edward C. Cocking, interview with author, 24 March 2016.

90. Cocking, "Plant Protoplasts" (2000), 80.

91. Shepard et al., "Genetic Transfer in Plants," 683. The phenomenon of chromosome segregation also occurs in animal cell fusion.

92. Power et al., "Some Consequences of the Fusion," 198.

93. Power et al., 206.

94. Shepard et al., "Genetic Transfer in Plants," 687.

95. Evans, "Somatic Hybrids for Crop Improvement," 857.

96. Evans, "Agricultural Applications of Plant Protoplast Fusion," 259.

97. Evans, "Somatic Hybrids for Crop Improvement," 858.

98. Cocking and Davey, "Gene Transfer in Cereals," 1262.

99. Pental and Cocking, "Some Theoretical and Practical Possibilities," 90.

100. Cocking, "Opportunities from the Use of Protoplasts," 564–65.

101. Cocking and Davey, "Gene Transfer in Cereals," 1262.

102. Cocking, "Opportunities from the Use of Protoplasts," 566.

103. Curry, "Industrial Evolution," 747.

104. "Cell Fusion," 1039.

105. Comar, "La Biotechnologie n'est plus ce qu'elle était."

106. Bud, "Biotechnology in the Twentieth Century," 416–17.

Conclusion: A Modern Rediscovery

1. Stegemann and Bock, "Exchange of Genetic Material between Cells," 650.

2. Stegemann and Bock, 651.

3. Fuentes et al., "Horizontal Genome Transfer as an Asexual Path," 233.

4. Fuentes et al., 235.

5. Gurdon et al., "Cell-to-Cell Movement of Mitochondria in Plants," 3397. Grafting may also be combined with gene editing. See Yang et al., "Heritable Transgene-Free Genome Editing in Plants."

6. Gurdon et al., "Cell-to-Cell Movement of Mitochondria in Plants," 3399.

7. On this distinction, see Ankeny and Leonelli, "What's So Special about Model Organisms?," 313–23.

8. On the former, see Charnley, "Experiments in Empire-Building," 292–300. On the latter, see Endersby, "Mutant Utopias," 471–503.

9. Magner, review of *Beyond the Gene*, 451–52.

10. Radick, *Disputed Inheritance*, 365.

11. Keller, *Century of the Gene*, 1–2.

12. Meloni, *Political Biology*, 32.

13. Meloni, 40.

14. Radick, *Disputed Inheritance*, 365–66.

15. Hodge, "Generation and the Origin of Species," 268.

16. Michurin, *Selected Works*, 200.

17. On breeders and near-utopian ambitions, see Curry, *Evolution Made to Order*; and Pawley, *Nature of the Future*.

18. Finlay, "China, the West, and World History," 273.

19. Mougey, "Needham at the Crossroads," 85.

20. Joseph Needham's notes on M. J. Hagerty's manuscript translation of a section "Thu Shu Chi Chhěng" [圖書集成 / Tu Shu Ji Cheng] (1726), relating to botany, especially citrus and grafting, 1950–1959, GBR/1928/NRI/SCC2/254/10/1, Joseph Needham Papers, Needham Research Institute, Cambridge.

21. Postcard from William Palmer of Homerton College, Cambridge, providing botanical references from "The *Georgics* of Virgil," 18 November 1965, GBR/1928/NRI/SCC2/254/10/3, Needham Papers.

22. Typescript translation of a passage by Chen Kuan [陳瓘 / Chen Guan] on grafting flowers, taken from the "Thu Shu Chi Chhěng" [圖書集成 / Tu Shu Ji Cheng] (1726), 1960–1969, GBR/1928/NRI/SCC2/254/10/5, Needham Papers.

23. Needham, Lu, and Huang, *Science and Civilisation in China*, 396.

24. Needham, Lu, and Huang, 643.

25. Victor Meally, letter to Joseph Needham, with a list of errata and newspaper cuttings, and copy of Needham's reply, 2 August 1984, GBR/1928/NRI/SCC4/11/24, Needham Papers.

26. On the contemporary historiographical importance of Needham, see Mei, "Some Reflections on Joseph Needham's Intellectual Heritage," 594–603.

27. Meloni, *Political Biology*, 140.

28. Popenoe, "Plant Chimeras," 521.

29. Hudson and Richens, *New Genetics in the Soviet Union*, 45–51.

30. Robert Pickard, in "GM Crops: Understanding the Issues," 2001, Box 1, Bob Fiddaman Collection of GM Material, Library and Archives, Science Museum, London.

31. See Bud, *Uses of Life*.

32. Pauly, *Fruits and Plains*.

33. Pauly, *Controlling Life*; Schneider, *Hybrid Nature*; and Sumner, *Brewing Science, Technology and Print*.

34. Dixon, "Putting the 'Bio' in Biotech," 38.

35. Historians have already moved to include household experimenters and gardeners in the development and promotion of biotechnology. See Curry, "From Garden Biotech to Garage Biotech," 539–65; Johnson, "Safeguarding the Atom," 551–71; and Holmes, "Houseflies and Fungi," 209–24.

36. On the tenuous link between biological innovation and intellectual property, see Olmstead and Rhode, *Creating Abundance*.

37. Liu, "Historical and Modern Genetics," 125.

38. Stegemann et al., "Horizontal Transfer of Chloroplast Genomes," 2437.

39. Fuentes et al., "Horizontal Genome Transfer as an Asexual Path," 235.

40. Gurdon et al., "Cell-to-Cell Movement of Mitochondria in Plants," 3397.

41. Le Page, "Farmers May Have Been Accidentally Making GMOs."

42. For the author's take on the GM controversy, see Holmes, "Perspectives on Biotechnology," 476–84.

43. Maienschein, "On Cloning," 423–24.

44. Maienschein, 431. On the need for historians to address current debates, see Guldi and Armitage, *History Manifesto*.

45. These hybrids were adopted by farmers partly for their superior characteristics: better yield and resilience. Yet wider political and economic factors were also at work behind the triumph of hybrid corn. See Fitzgerald, *Business of Breeding*.

46. Duvick, "Biotechnology in the 1930s," 72.

47. Graham, *What Have We Learned*, 19.

48. Radick, "Other Histories, Other Biologies," 33.

BIBLIOGRAPHY

Archives

Bateman, Angus John. Correspondence. This correspondence is incorporated into the J. B. S. Haldane Papers. University College London (UCL) Archives. Accessed through UCL Digital Collections. https://www.ucl.ac.uk/library/digital-collections.

Bateson, William. Papers. Special Collections, Cambridge University Library.

Bell, G. B. H. Correspondence. John Innes Centre, Norwich, England.

Darwin, Charles. Correspondence. Cambridge University Library. Accessed through the Charles Darwin Correspondence Project. https://www.darwinproject.ac.uk/.

Davenport, Charles Benedict. Papers. Library of the American Philosophical Society, Philadelphia.

Fiddaman, Bob. Collection of GM Material. Library and Archives, Science Museum, London.

Haldane, J. B. S. Papers. University College London (UCL) Archives. Accessed through UCL Digital Collections. https://www.ucl.ac.uk/library/digital-collections.

McLaren, Anne. Papers. British Library, London.

Needham, Joseph. Papers. Needham Research Institute, Cambridge.

Tansley, Arthur. Papers. Special Collections, Cambridge University Library.

Waddington, Conrad Hal. Papers. Special Collections, Edinburgh University Library.

Watson, James D. Collection. Archives Repository, Cold Spring Harbor Laboratory.

Weiss, Frederick Ernest. Correspondence. Linnean Society Archives, London.

Books and Articles

Agar, Jon. "What Happened in the Sixties?" *British Journal for the History of Science* 41, no. 4 (2008): 567–600.

Alberti, Samuel J. M. M. "Amateurs and Professionals in One County: Biology and Natural History in Late Victorian Yorkshire." *Journal of the History of Biology* 34, no. 1 (2001): 115–47.

Allen, Garland E. "Hugo de Vries and the Reception of the 'Mutation Theory.'" *Journal of the History of Biology* 2, no. 1 (1969): 55–87.

Allen, Garland E. "Origins of the Classical Gene Concept, 1900–1950: Genetics, Mechanistic, Philosophy, and the Capitalization of Agriculture." *Perspectives in Biology and Medicine* 57, no. 1 (2014): 8–39.

Ankeny, Rachel A., and Sabina Leonelli. "What's So Special about Model Organisms?" *Studies in History and Philosophy of Science* 42, no. 2 (2011): 313–23.

Bailey, L. H. *Cyclopedia of American Horticulture.* 5th ed. Vol. 2. New York: Macmillan, 1906.

Bateman, A. J. "Grafting Experiments between the Tomato Varieties, Golden Apple and Oxheart." *Nature* 175, no. 4469 (1955): 1118–20.

Bateson, William. "Dr. Kammerer's Testimony to the Inheritance of Acquired Characters." *Nature* 103, no. 2592 (1919): 344–45.

Bateson, W[illiam]. "Hybridisation and Cross-Breeding as a Method of Scientific Investigation." *Journal of the Royal Horticultural Society* 24 (1900): 59–66.

Bateson, William. *Mendel's Principles of Heredity.* Cambridge: Cambridge University Press, 1909.

Bateson, W[illiam]. "Practical Aspects of the New Discoveries in Heredity." *Memoirs of the Horticultural Society of New York* 1 (1902): 1–9.

Bateson, William. *Problems of Genetics.* New Haven, CT: Yale University Press, 1913.

Bateson, William. "Root-Cuttings, Chimaeras and 'Sports,'" *Journal of Genetics* 6, no. 2 (1916): 75–80.

Bateson, William, and Beatrice Bateson. *Letters from the Steppe: Written in the Years 1886–1887.* London: Methuen, 1928.

Bengtsson, Bengt O., and Anna Tunlid. "The 1948 International Congress of Genetics in Sweden: People and Politics." *Genetics* 185, no. 3 (2010): 709–15.

Bevan, Michael W., Richard B. Flavell, and Mary-Dell Chilton. "A Chimaeric Antibiotic Resistance Gene as a Selectable Marker for Plant Cell Transformation." *Nature* 304 (1983): 184–87.

Bhojwani, S. S., P. K. Evans, and E. C. Cocking. "Protoplast Technology in Relation to Crop Plants: Progress and Problems." *Euphytica* 26 (1977): 343–60.

Billingham, R. E., L. Brent, and P. B. Medawar. "'Actively Acquired Tolerance' of Foreign Cells." *Nature* 172, no. 4379 (1953): 603–6.

Billingham, R. E., L. Brent, and P. B. Medawar. "Quantitative Studies on Tissue Transplantation Immunity. III. Actively Acquired Tolerance." *Philosophical Transactions of the Royal Society of London. Series B, Biological Sciences* 239, no. 666 (1956): 357–414.

Blackman, V. H. "Some Recent Work on Hybrids in Plants." *New Phytologist* 1, no. 4 (1902): 73–80.

Bonneuil, Christophe. "Mendelism, Plant Breeding and Experimental Cultures: Agriculture and the Development of Genetics in France." *Journal of the History of Biology* 39, no. 2 (2006): 281–308.

Bonneuil, Christophe. "Pure Lines as Industrial Simulacra: A Cultural History of Genetics from Darwin to Johannsen." In *Heredity Explored: Between Public Domain and Experimental Science, 1850–1930,* edited by Staffan Müller-Wille and Christina Brandt, 213–42. Cambridge, MA: MIT Press, 2016.

Boulter, Michael. *Bloomsbury Scientists: Science and Art in the Wake of Darwin.* London: UCL Press, 2017.

Boveri, Theodor. "Concerning the Origin of Malignant Tumours." Translated and annotated by Henry Harris. *Journal of Cell Science* 121, suppl. 1 (2008): 1–84.

Bowler, Peter J. *The Eclipse of Darwinism: Anti-Darwinian Evolution Theories in the Decades around 1900.* Baltimore, MD: Johns Hopkins University Press, 1983.

Bowler, Peter J. *The Mendelian Revolution: The Emergence of Hereditarian Concepts in Modern Science and Society.* London: Athlone, 1989.

Brandt, Christina. "Development and Heredity in the Interwar Period: Hans Spemann and Fritz Baltzer on Organizers and Merogones." *Journal of the History of Biology* 55, no. 2 (2022): 253–83.

Brannigan, Augustine. "The Reification of Mendel." *Social Studies of Science* 9, no. 4 (1979): 423–54.

Brent, Leslie. *A History of Transplantation Immunology.* San Diego, CA: Academic Press, 1997.

Browne, Janet. *Charles Darwin: The Power of Place.* Vol. 2. Princeton, NJ: Princeton University Press, 2003.

Brush, Stephen G. *Making 20th Century Science: How Theories Became Knowledge.* New York: Oxford University Press, 2015.

Bud, Robert. "Biotechnology in the Twentieth Century." *Social Studies of Science* 21, no. 3 (1991): 415–57.

Bud, Robert. *The Uses of Life: A History of Biotechnology.* Cambridge: Cambridge University Press, 1993.

Buffum, B. C. "Effect of Environment on Plant Breeding." *Journal of Heredity*, o.s., 6, no. 1 (1911): 212–24.

Burbank, Luther. "Heredity." *Journal of Heredity*, o.s., 1, no. 2 (1905): 158–61.

Burian, Richard M., and Doris T. Zallen. "Genes." In *The Cambridge History of Science*, vol. 6, *The Modern Biological and Earth Sciences*, edited by P. J. Bowler and J. V. Pickstone, 432–50. Cambridge: Cambridge University Press, 2009.

Burkhardt, Richard W., Jr. "Lamarck, Cuvier, and Darwin on Animal Behavior and Acquired Characters." In *Transformations of Lamarckism: From Subtle Fluids to Molecular Biology*, edited by Snait B. Gissis and Eva Jablonka, 33–44. Cambridge, MA: MIT Press, 2011.

Burns, Marca. *The Genetics of the Dog*. Edinburgh: Commonwealth Bureau of Animal Breeding and Genetics, 1952.

Calver, Neil. "Sir Peter Medawar: Science, Creativity and the Popularization of Karl Popper." *Notes and Records of the Royal Society* 67, no. 4 (2013): 301–14.

Campbell, Christopher. *The Botanist and the Vintner: How Wine Was Saved for the World*. Chapel Hill, NC: Algonquin Books, 2005.

Campbell, Douglas Houghton. "The Nature of Graft-Hybrids." *American Naturalist* 45, no. 529 (1911): 41–53.

Campbell, Douglas Houghton. *Plant Life and Evolution*. New York: Henry Holt, 1911.

Campos, Luis. "Dialectics Denied: Muller, Lysenkoism, and the Fate of Chromosomal Mutation." In *The Lysenko Controversy as a Global Phenomenon*, edited by William deJong-Lambert and Nikolai Krementsov, vol. 2, *Genetics and Agriculture in the Soviet Union and Beyond*, 161–84. Cham, Switzerland: Palgrave Macmillan, 2017.

Campos, Luis. "Genetics without Genes: Blakeslee, Datura, and 'Chromosomal Mutations.'" In *Conference: Heredity in the Century of the Gene (A Cultural History of Heredity IV)*, preprint 343, 243–57. Berlin: Max Planck Institute for the History of Science, 2008.

Carleton, M. A. "Fundamental Requirements for Grain Breeding." *Journal of Heredity*, o.s., 2, no. 1 (1906): 129–35.

Carlson, Elof Axel. *Genes, Radiation, and Society: The Life and Work of H. J. Muller*. Ithaca, NY: Cornell University Press, 1981.

Carlson, Peter S., Harold H. Smith, and Rosemarie D. Dearing. "Parasexual Interspecific Plant Hybridization." *Proceedings of the National Academy of Sciences* 69, no. 8 (1972): 2292–94.

Casselton, Lorna A., and David A. Jones. "Dan Lewis. 30 December 1910–30 September 2009." *Biographical Memoirs of Fellows of the Royal Society* 58 (2012): 163–78.

Castle, W. E. "An Apple Chimera." *Journal of Heredity* 5, no. 5 (1914): 200–202.

Castle, W. E. "Mendel's Law of Heredity." *Proceedings of the American Academy of Arts and Sciences* 38, no. 18 (1903): 535–48.

Castle, W. E. "On 'Soma Influence' in Ovarian Transplantation." *Science*, n.s., 34, no. 865 (1911): 113–15.

Castle, W. E., and John C. Phillips. *On Germinal Transplantation in Vertebrates*. Washington, DC: Carnegie Institution of Washington, 1911.

Castle, W. E., and John C. Phillips. "A Successful Ovarian Transplantation in the Guinea-Pig, and Its Bearing on Problems of Genetics." *Science* 30, no. 766 (1909): 312–13.

"Cell Fusion: A New Gift to Biology." *Nature* 223, no. 5210 (1969): 1039–41.

Chadarevian, Soraya de. *Heredity under the Microscope: Chromosomes and the Study of the Human Genome*. Chicago: University of Chicago Press, 2020.

Charles, Daniel. *Lords of the Harvest: Biotech, Big Money, and the Future of Food*. Cambridge, MA: Perseus, 2001.

Charnley, Berris. "Experiments in Empire-Building: Mendelian Genetics as a National, Imperial, and Global Agricultural Enterprise." *Studies in History and Philosophy of Science Part A* 44, no. 2 (2013): 292–300.

Chittenden, Reginald J. "Vegetative Segregation." *Bibliographia Genetica* 3 (1927): 355–439.

Churchill, Frederick B. *August Weismann: Development, Heredity, and Evolution.* Cambridge, MA: Harvard University Press, 2015.

Clark, Ronald. *J.B.S.: The Life and Work of J. B. S. Haldane.* Oxford: Oxford University Press, 1968.

Clowes, F. A. L. Review of *Plant Chimeras*, by W. Neilson-Jones. *New Phytologist* 68, no. 4 (1969): 1252–53.

Cock, Alan G., and Donald R. Forsdyke. *Treasure Your Exceptions: The Science and Life of William Bateson.* New York: Springer, 2008.

Cocking, E. C. "Opportunities from the Use of Protoplasts." *Philosophical Transactions of the Royal Society of London. Series B, Biological Sciences* 292, no. 1062 (1981): 557–68.

Cocking, E. C. "Plant Protoplasts." In *Viewpoints in Biology*, vol. 4, edited by J. D. Cathy and C. L. Duddington, 170–203. London: Butterworths, 1965.

Cocking, Edward C. "Plant Protoplasts." *In Vitro Cellular & Developmental Biology. Plant* 36, no. 2 (2000): 77–82.

Cocking, E. C. "Properties of Isolated Plant Protoplasts." *Nature* 191, no. 4790 (1961): 780–82.

Cocking, E. C. Review of *Protoplast Fusion: Genetic Engineering in Higher Plants*, by Y. Y. Gleba and K. M. Sytnik. *Heredity* 57 (1986): 432.

Cocking, Edward C., and Michael R. Davey. "Gene Transfer in Cereals." *Science* 236, no. 4806 (1987): 1259–62.

Cocking, E. C., D. George, M. J. Price-Jones, and J. B. Power. "Selection Procedures for the Production of Inter-species Somatic Hybrids of *Petunia hybrida* and *Petunia parodii* II. Albino Complementation Selection." *Plant Science Letters* 10, no. 1 (1977): 7–12.

Cocking, E. C., and D. W. Gregory. "Organized Protoplasmic Units of the Plant Cell. I. Their Occurrence, Origin, and Structure." *Journal of Experimental Botany* 14, no. 3 (1963): 504–11.

Coen, Deborah R. "Living Precisely in Fin-de-Siècle Vienna." *Journal of the History of Biology* 39, no. 3 (2006): 493–523.

Collins, H. M. "Son of Seven Sexes: The Social Destruction of a Physical Phenomenon." *Social Studies of Science* 11, no. 1 (1981): 33–62.

Comar, Jean. "La Biotechnologie n'est plus ce qu'elle était." *Biofutur*, December 1986, 5.

Constabel, F. "Somatic Hybridization in Higher Plants." *In Vitro* 12, no. 11 (1976): 743–48.

Cook, Robert C. "Bacon Predicted Triumphs of Plant Breeding." *Journal of Heredity* 23, no. 4 (1932): 162–65.

Cook, Robert C. "Lysenko's Marxist Genetics: Science or Religion?" *Journal of Heredity* 40, no. 7 (1949): 169–202.

Cornforth, Maurice. *Communism and Philosophy: Contemporary Dogmas and Revisions of Marxism.* London: Lawrence and Wishart, 1980.

"A Correspondent." *Gardeners' Chronicle* 3 (1842): 35.

Corsi, Pietro. "Jean-Baptiste Lamarck: From Myth to History." In *Transformations of Lamarckism: From Subtle Fluids to Molecular Biology*, edited by Snait B. Gissis and Eva Jablonka, 9–18. Cambridge, MA: MIT Press, 2011.

Cowles, Henry C., and Charles J. Chamberlain. "Graft Hybrids and Chimeras." *Botanical Gazette* 51, no. 2 (1911): 147–53.

Crane, M. B. Review of *Selected Works*, by I. V. Michurin. *Journal of the Royal Horticultural Society* 75 (1950): 369–70.

Creese, Mary R. S., and Thomas M. Creese. "British Women Who Contributed to Research in the Geological Sciences in the Nineteenth Century." *British Journal for the History of Science* 27, no. 1 (1994): 23–54.

Crew, F. A. E. *Animal Genetics: An Introduction to the Science of Animal Breeding*. Edinburgh: Oliver and Boyd, 1925.

Crowe, Nathan. *Forgotten Clones: The Birth of Cloning and the Biological Revolution*. Pittsburgh: University of Pittsburgh Press, 2021.

Cunningham, J. T. "Breeding Experiments on the Inheritance of Acquired Characters." *Nature* 111, no. 2795 (1923): 702.

Curry, Helen Anne. *Evolution Made to Order: Plant Breeding and Technological Innovation in Twentieth-Century America*. Chicago: University of Chicago Press, 2016.

Curry, Helen Anne. "From Garden Biotech to Garage Biotech: Amateur Experimental Biology in Historical Perspective." *British Journal for the History of Science* 47, no. 3 (2014): 539–65.

Curry, Helen Anne. "Industrial Evolution: Mechanical and Biological Innovation at the General Electric Research Laboratory." *Technology and Culture* 54, no. 4 (2013): 746–81.

Darlington, C. D. Review of *The Genetics of the Dog*, by Marca Burns. *Heredity* 7 (1953): 141–42.

Darwin, Charles. *The Variation of Animals and Plants under Domestication*, Vol. 1. London: John Murray, 1868.

Davenport, Charles B. *Inheritance of Characteristics in Domestic Fowl*. Washington, DC: Carnegie Institution of Washington, 1909.

Davenport, C. B. "Mendel's Law of Dichotomy in Hybrids." *Biological Bulletin* 2, no. 6 (1901): 307–10.

Davenport, C. B. "The Transplantation of Ovaries in Chickens." *Journal of Morphology* 22, no. 1 (1911): 111–22.

Davenport, Chas. B. "Zoology of the Twentieth Century." *Science* 14, no. 348 (1901): 315–24.

deJong-Lambert, William. *The Cold War Politics of Genetic Research: An Introduction to the Lysenko Affair*. Dordrecht: Springer, 2012.

deJong-Lambert, William. "H. J. Muller and J. B. S. Haldane: Eugenics and Lysenkoism." In *The Lysenko Controversy as a Global Phenomenon*, edited by William deJong-Lambert and Nikolai Krementsov, vol. 2, *Genetics and Agriculture in the Soviet Union and Beyond*, 103–35. Cham, Switzerland: Palgrave Macmillan, 2017.

Dewsbury, Donald A. "The Darwin-Bateman Paradigm in Historical Context." *Integrative and Comparative Biology* 45, no. 5 (2005): 831–37.

Dierig, Sven, Jens Lachmund, and J. Andrew Mendelsohn. "Introduction: Toward an Urban History of Science." In *Science and the City*, edited by Sven Dierig, Jens Lachmund, and J. Andrew Mendelsohn, vol. 18 of *Osiris* (2003): 1–19.

Dixon, Bernard. "Putting the 'Bio' in Biotech." *New Scientist*, January 31, 1985, 38.

Duančić, Vedran. "Lysenko in Yugoslavia, 1945–1950s: How to De-Stalinize Stalinist Science." *Journal of the History of Biology* 53, no. 1 (2020): 159–94.

Dunn, L. C. "William Ernest Castle, 1867–1962." *Biographical Memoirs of the National Academy of Sciences* 38 (1965): 33–80.

Duvick, Donald N. "Biotechnology in the 1930s: The Development of Hybrid Maize." *Nature Reviews Genetics* 2 (2001): 69–74.

Elina, Olga, Susanne Heim, and Nils Roll-Hansen. "Plant Breeding on the Front: Imperialism, War, and Exploitation." In *Politics and Science in Wartime: Comparative International Perspectives on the Kaiser Wilhelm Institute*, edited by Carola Sachse and Mark Walker, vol. 20 of *Osiris* (2005): 161–79.

Endersby, Jim. "Mutant Utopias: Evening Primroses and Imagined Futures in Early Twentieth-Century America." *Isis* 104, no. 3 (2013): 471–503.

Evans, David A. "Agricultural Applications of Plant Protoplast Fusion." *Bio/Technology* 1, no. 3 (1983): 253–61.

Evans, David A. "Somatic Hybrids for Crop Improvement and Gene Research." *Bio/Technology* 1, no. 10 (1983): 856–58.

Felföldy, L. J. M. "Tomato Grafting Experiments at Tihany." *Nature* 179, no. 4570 (1957): 1144.

Finlay, Robert. "China, the West, and World History in Joseph Needham's *Science and Civilisation in China*." *Journal of World History* 11, no. 2 (2000): 265–303.

Fitzgerald, Deborah. *The Business of Breeding: Hybrid Corn in Illinois, 1890–1940*. Ithaca, NY: Cornell University Press, 1990.

Fitzgerald, Deborah. *Every Farm a Factory: The Industrial Ideal in American Agriculture*. New Haven, CT: Yale University Press, 2003.

Fitzgerald, Deborah. "Farmers Deskilled: Hybrid Corn and Farmers' Work." *Technology and Culture* 34, no. 2 (1993): 324–43.

Fleming, Donald. "On Living in a Biological Revolution." *Atlantic Monthly*, February 1969, 64–70.

Fletcher, Harold Roy. *The Story of the Royal Horticultural Society, 1804–1968*. London: Oxford University Press, 1969.

Forrester, John, and Laura Cameron. *Freud in Cambridge*. Cambridge: Cambridge University Press, 2017.

Foster, Toshi M., and Maria José Aranzana. "Attention Sports Fans! The Far-Reaching Contributions of Bud Sport Mutants to Horticulture and Plant Biology." *Horticulture Research* 5, no. 44 (2018): 1–13.

Fowler, M. W. "Plant Cell Biotechnology and Agriculture: Impacts and Perspectives." *Philosophical Transactions of the Royal Society of London. Series B, Biological Sciences* 310, no. 1144 (1985): 215–20.

Frank, Margaret H., and Daniel H. Chitwood. "Plant Chimeras: The Good, the Bad, and the 'Bizzaria,'" *Developmental Biology* 419, no. 1 (2016): 41–53.

Franklin, Sarah. "Obituary: Dame Dr Anne McLaren." *Regenerative Medicine* 2, no. 5 (2007): 853–59.

Frey-Wyssling, A. "Gymnoplasts Instead of 'Protoplasts.'" *Nature* 216, no. 5114 (1967): 516.

Friedman, David M. *The Immortalists: Charles Lindbergh, Dr. Alexis Carrel, and Their Daring Quest to Live Forever*. New York: Ecco, 2007.

Fuentes, Ignacia, Sandra Stegemann, Hieronim Golczyk, Daniel Karcher, and Ralph Bock. "Horizontal Genome Transfer as an Asexual Path to the Formation of New Species." *Nature* 511, no. 7508 (2014): 232–35.

Galera, Andrés. "The Impact of Lamarck's Theory of Evolution before Darwin's Theory." *Journal of the History of Biology* 50, no. 1 (2017): 53–70.

Galston, Arthur W. "Molecular Biology and Agricultural Botany." In *The Social Impact of Modern Biology*, edited by Watson Fuller, 154–66. London: Routledge, 1971.

Galton, Francis. "Experiments in Pangenesis, by Breeding from Rabbits of a Pure Variety, into Whose Circulation Blood Taken from Other Varieties had Previously Been Largely Transfused." *Proceedings of the Royal Society of London* 19 (1871): 393–410.

Gates, R. R. "Graft Hybrids." *Botanical Gazette* 47, no. 1 (1909): 84.

Gates, R. R. "Graft Hybrids." *Botanical Gazette* 47, no. 3 (1909): 250.

Gates, R. R. "Graft Hybrids." *Botanical Gazette* 48, no. 6 (1909): 478.

Gates, R. R. "Graft Hybrids." *Botanical Gazette* 49, no. 5 (1910): 386–87.

Gates, R. R. "Mendelism." *Botanical Gazette* 48, no. 1 (1909): 61–62.

Geison, Gerald L. "Darwin and Heredity: The Evolution of His Hypothesis of Pangenesis." *Journal of the History of Medicine and Allied Sciences* 24, no. 4 (1969): 375–411.

Gleba, Y. Y., K. M. Sytnik, and R. L. Shoeman, eds. *Protoplast Fusion: Genetic Engineering in Higher Plants*. Berlin: Springer, 1984.

Gliboff, Sander. "The Case of Paul Kammerer: Evolution and Experimentation in the Early 20th Century." *Journal of the History of Biology* 39, no. 3 (2006): 525–63.

Gliboff, Sander. "The Golden Age of Lamarckism, 1866–1926." In *Transformations of Lamarckism: From Subtle Fluids to Molecular Biology*, edited by Snait B. Gissis and Eva Jablonka, 45–55. Cambridge, MA: MIT Press, 2011.

Goncharov, N. P., and N. I. Savel'ev. "Ivan V. Michurin: On the 160th Anniversary of the Birth of the Russian Burbank." *Russian Journal of Genetics: Applied Research* 6, no. 1 (2016): 105–27.

Gordin, Michael D. "How Lysenkoism Became Pseudoscience: Dobzhansky to Velikovsky." *Journal of the History of Biology* 45, no. 3 (2012): 443–68.

Gordin, Michael D. "Lysenko Unemployed: Soviet Genetics after the Aftermath." *Isis* 109, no. 1 (2018): 56–78.

Gould, Stephen Jay. *Hen's Teeth and Horse's Toes*. New York: Norton, 1983.

"Graft-Inheritance." *Nature* 113, no. 2831 (1924): 174.

Graham, Loren R. *Lysenko's Ghost: Epigenetics and Russia*. Cambridge, MA: Harvard University Press, 2016.

Graham, Loren R. *Science and Philosophy in the Soviet Union*. New York: Knopf, 1972.

Graham, Loren R. *What Have We Learned about Science and Technology from the Russian Experience?* Stanford, CA: Stanford University Press, 1998.

Groner, Yoram, Pnina Sachs, and Joseph Lotem. "Leo Sachs. 14 October 1924–12 December 2013." *Biographical Memoirs of Fellows of the Royal Society* 66 (2019): 355–75.

Guldi, Jo, and David Armitage. *The History Manifesto*. Cambridge: Cambridge University Press, 2014.

Gurdon, Csanad, Zora Svab, Yaping Feng, Dibyendu Kumar, and Pal Maliga. "Cell-to-Cell Movement of Mitochondria in Plants." *Proceedings of the National Academy of Sciences* 113, no. 12 (2016): 3395–400.

Guthrie, C. C. "On Evidence of Soma Influence on Offspring from Engrafted Ovarian Tissue." *Science*, n.s., 33, no. 856 (1911): 816–19.

Guthrie, C. C. "Further Results of Transplantation of Ovaries in Chickens." *Journal of Experimental Zoology* 5, no. 4 (1908): 563–76.

Guthrie, C. C. "On Graft Hybrids." *Journal of Heredity*, o.s., 6, no. 1 (1911): 356–73.

Guthrie, C. C. "Guinea Pig Graft-Hybrids." *Science* 30, no. 777 (1909): 724–25.

Guthrie, C. C. "Transplantation of Ovaries." *Science*, n.s., 34, no. 887 (1911): 918.

Haines, Catharine M. C. *International Women in Science: A Biographical Dictionary to 1950*. Santa Barbara, CA: ABC-CLIO, 2001.

Haldane, J. B. S. *Science Advances*. London: George Allen & Unwin, 1947.

Haldane, J. B. S. "Some Alternatives to Sex." *New Biology* 19 (1955): 7–26.

Hamilton, David. *The First Transplant Surgeon: The Flawed Genius of Nobel Prize Winner, Alexis Carrel*. Hackensack, NJ: World Scientific, 2017.

Hamilton, David. *A History of Organ Transplantation: Ancient Legends to Modern Practice*. Pittsburgh: University of Pittsburgh Press, 2012.

Hansen, Niels Ebbesen. "Methods in Breeding Hardy Fruits." *Journal of Heredity*, o.s., 2, no. 1 (1906): 168–69.

Harman, Oren Solomon. "C. D. Darlington and the British and American Reaction to Lysenko and the Soviet Conception of Science." *Journal of the History of Biology* 36, no. 2 (2003): 309–52.

Harman, Oren Solomon. *The Man Who Invented the Chromosome: A Life of Cyril Darlington.* Cambridge, MA: Harvard University Press, 2004.

Harris, Henry. *The Cells of the Body: A History of Somatic Cell Genetics.* New York: Cold Spring Harbor Laboratory Press, 1995.

Harris, Henry. "Review Lecture: Hybrid Cells from Mouse and Man: A Study in Genetic Regulation." *Proceedings of the Royal Society of London. Series B, Biological Sciences* 166, no. 1004 (1966): 358–68.

Harris, Henry, J. F. Watkins, G. L. Campbell, E. P. Evans, and C. E. Ford. "Mitosis in Hybrid Cells Derived from Mouse and Man." *Nature* 207, no. 4997 (1965): 606–8.

Harvey, R. D. "Pioneers of Genetics: A Comparison of the Attitudes of William Bateson and Erwin Baur to Eugenics." *Notes and Records of the Royal Society* 49, no. 1 (1995): 105–17.

Harwood, Jonathan. "Did Mendelism Transform Plant Breeding? Genetic Theory and Breeding Practice, 1900–1945." In *New Perspectives on the History of Life Sciences and Agriculture,* edited by Denise Phillips and Sharon Kingsland, 345–70. Cham, Switzerland: Springer, 2015.

Harwood, Jonathan. *Styles of Scientific Thought: The German Genetics Community, 1900–1933.* Chicago: University of Chicago Press, 1993.

Hašek, M. "Tolerance Phenomena in Birds." *Proceedings of the Royal Society of London. Series B, Biological Sciences* 146, no. 922 (1956): 67–77.

Heape, Walter. "Preliminary Note on the Transplantation and Growth of Mammalian Ova within a Uterine Foster-Mother." *Proceedings of the Royal Society* 48, no. 292–95 (1891): 457–58.

Herbert, W. "Further Remarks on the Cytisus Adami." *Gardener's Magazine,* n.s., 6 (1840): 381–83.

Herbert, W. "On the Singular Origin of the Purple Laburnum." *Gardener's Magazine,* n.s., 6 (1840): 289–90.

Hodge, M. J. S. "Generation and the Origin of Species (1837–1937): A Historiographical Suggestion." *British Journal for the History of Science* 22, no. 3 (1989): 267–81.

Hogan, Brigid. "From Embryo to Ethics: A Career in Science and Social Responsibility: An Interview with Anne McLaren." *International Journal of Developmental Biology* 45 (2001): 477–82.

Holmes, Matthew. "Changing Techniques in Crop Plant Classification: Molecularization at the National Institute of Agricultural Botany during the 1980s." *Annals of Science* 74, no. 2 (2017): 149–64.

Holmes, Matthew. "Crops in a Machine: Industrialising Barley Breeding in Twentieth-Century Britain." In *Histories of Technology, the Environment and Modern Britain,* edited by Jon Agar and Jacob Ward, 142–60. London: UCL Press, 2018.

Holmes, Matthew. "Houseflies and Fungi: The Promise of an Early Twentieth-Century Biotechnology." *Notes and Records* 76, no. 1 (2022): 209–24.

Holmes, Matthew. "Perspectives on Biotechnology: Public and Corporate Narratives in the GM Archives." *Plants, People, Planet* 4, no. 5 (2022): 476–84.

Holmes, Matthew. "Somatic Hybridization: The Rise and Fall of a Mid-Twentieth-Century Biotechnology." *Historical Studies in the Natural Sciences* 48, no. 1 (2018): 1–23.

Holmes, Matthew. "Yeast, Coal, and Straw: J. B. S. Haldane's Vision for the Future of Science and Synthetic Food." *History of the Human Sciences* 36, no. 3–4 (2023): 202–20.

Holterhoff, Kate. "The History and Reception of Charles Darwin's Hypothesis of Pangenesis." *Journal of the History of Biology* 47, no. 4 (2014): 661–95.

Howard, H. W. Review of *Plant Chimeras,* by W. Neilson-Jones. *Heredity* 24 (1969): 503–4.

Hudson, P. S., and R. H. Richens. *The New Genetics in the Soviet Union.* Cambridge: School of Agriculture, 1946.

Hughes, Hilary M. "Modern Techniques in Fruit Growing." *Journal of the Royal Horticultural Society* 96 (1971): 222–26.

Hughes, Sally Smith. "Making Dollars Out of DNA: The First Major Patent in Biotechnology and the Commercialization of Molecular Biology, 1974–1980." *Isis* 92, no. 3 (2001): 541–75.

Hume, Margaret. "On the Presence of Connecting Threads in Graft Hybrids." *New Phytologist* 12, no. 6 (1913): 216–21.

Huxley, Julian. *Soviet Genetics and World Science: Lysenko and the Meaning of Heredity.* London: Chatto and Windus, 1949.

Hyun, Insoo. *Bioethics and the Future of Stem Cell Research.* Cambridge: Cambridge University Press, 2013.

Iida, Kaori. "A Controversial Idea as a Cultural Resource: The Lysenko Controversy and Discussions of Genetics as a 'Democratic' Science in Postwar Japan." *Social Studies of Science* 45, no. 4 (2015): 546–69.

Illman, John. "Leslie Brent: Junior Member of the 'Holy Trinity of Immunology,'" *BMJ* 368 (2020), https://doi.org/10.1136/bmj.m977.

Ivanyi, Juraj. "Milan Hašek and the Discovery of Immunological Tolerance." *Nature Reviews Immunology* 3, no. 7 (2003): 591–97.

Janaki Ammal, E. K. "Chromosomes and Horticulture." *Journal of the Royal Horticultural Society* 76 (1951): 236–39.

Jinks, John L. *Extrachromosomal Inheritance.* London: Prentice-Hall, 1964.

Johannsen, W. "Inheritance of Characters Acquired by Grafting." *Nature* 113, no. 2841 (1924): 536.

Johnson, Paige. "Safeguarding the Atom: The Nuclear Enthusiasm of Muriel Howorth." *British Journal for the History of Science* 45, no. 4 (2012): 551–71.

Joravsky, David. *The Lysenko Affair.* Chicago: University of Chicago Press, 1986.

Joravsky, David. *Soviet Marxism and Natural Science, 1917–1932.* New York: Routledge, 2009. First published 1961 by Routledge and Kegan Paul.

Kammerer, Paul. "Breeding Experiments on the Inheritance of Acquired Characters." *Nature* 111, no. 2793 (1923): 637–40.

Kammerer, Paul. *The Inheritance of Acquired Characteristics.* New York: Boni and Liveright, 1924.

Keller, Evelyn Fox. *The Century of the Gene.* Cambridge, MA: Harvard University Press, 2000.

Keller, W. A., R. S. Pandeya, S. C. Gleddie, and G. Setterfield. "Application of Somatic Hybridization Technology to Plant Breeding." In *Genetic Manipulation in Crops: Proceedings of the International Symposium . . . Beijing, October 1984,* edited by the International Rice Research Institute, 192–93. London: Cassell Tycooly, 1988.

Kenney, Martin. *Biotechnology: The University–Industrial Complex.* New Haven, CT: Yale University Press, 1986.

Kihara, Hitoshi, and Karl Sax. "Genetics in the U.S.S.R." *Journal of Heredity* 44, no. 4 (1953): 132–58.

Kimmelman, Barbara A. "The American Breeders' Association: Genetics and Eugenics in an Agricultural Context, 1903–13." *Social Studies of Science* 13, no. 2 (1983): 163–204.

Kingsbury, Noel. *Hybrid: The History and Science of Plant Breeding.* Chicago: University of Chicago Press, 2009.

Klein, George, and Eva Klein. "How One Thing Has Led to Another." *Annual Review of Immunology* 7 (1989): 1–33.

Kleinman, Daniel Lee. *Impure Cultures: University Biology and the World of Commerce.* Madison: University of Wisconsin Press, 2003.

Klercker, John. "A Method for the Isolation of Living Protoplasts." *Plant Physiological Releases* 3 (1892): 463–74.

Kloppenburg, Jack Ralph, Jr. *First the Seed: The Political Economy of Plant Biotechnology, 1492–2000.* Cambridge: Cambridge University Press, 1988.

Knight, Robert L. *Abstract Bibliography of Fruit Breeding and Genetics to 1960: Malus and Pyrus.* Farnham Royal: Commonwealth Agricultural Bureaux, 1963.

Koestler, Arthur. *The Case of the Midwife Toad.* London: Hutchinson, 1971.

Kohler, Robert E. *Lords of the Fly: Drosophila Genetics and the Experimental Life.* Chicago: University of Chicago Press, 1994.

Kolchinsky, Eduard I. "Current Attempts at Exonerating 'Lysenkoism' and Their Causes." In *The Lysenko Controversy as a Global Phenomenon,* edited by William deJong-Lambert and Nikolai Krementsov, vol. 2, *Genetics and Agriculture in the Soviet Union and Beyond,* 207–36. Cham, Switzerland: Palgrave Macmillan, 2017.

Krementsov, Nikolai. "A 'Second Front' in Soviet Genetics: The International Dimension of the Lysenko Controversy, 1944–1947." *Journal of the History of Biology* 29, no. 2 (1996): 229–50.

Krementsov, Nikolai. *Stalinist Science.* Princeton, NJ: Princeton University Press, 1997.

Küster, Ernst. "Über die Gewinnung nackter Protoplasten." *Protoplasma* 3 (1927): 223–34.

Landecker, Hannah. *Culturing Life: How Cells Became Technologies.* Cambridge, MA: Harvard University Press, 2007.

Le Page, Michael. "Farmers May Have Been Accidentally Making GMOs for Millennia." *New Scientist,* March 7, 2016, https://www.newscientist.com/article/2079813-farmers-may-have-been-accidentally-making-gmos-for-millennia/.

Li Xianghui. "Advances in Plant Genetic Manipulation." In *Genetic Manipulation in Crops: Proceedings of the International Symposium . . . Beijing, October 1984,* edited by the International Rice Research Institute, 219–20. London: Cassell Tycooly, 1988.

Lidwell-Durnin, John. "The Production of a Physiological Puzzle: How *Cytisus adami* Confused and Inspired a Century's Botanists, Gardeners, and Evolutionists." *History and Philosophy of the Life Sciences* 40, no. 3 (2018): 1–22.

Lindegren, Carl C. *The Cold War in Biology.* Ann Arbor, MI: Planarian, 1966.

Liu, Yongsheng. "Historical and Modern Genetics of Plant Graft Hybridization." *Advances in Genetics* 56 (2006): 101–29.

Liu, Yongsheng. "Expanding the Potential of Plant Interfamily Grafting." Letter to the editor. *Plant and Cell Physiology* 63, no. 4 (2022): 448–49.

Logan, Cheryl A. *Hormones, Heredity, and Race: Spectacular Failure in Interwar Vienna.* New Brunswick, NJ: Rutgers University Press, 2013.

Loison, Laurent. "French Roots of French Neo-Lamarckisms, 1879–1985." *Journal of the History of Biology* 44, no. 4 (2011): 713–44.

Lönnig, W.-E., and H. Saedler. "Baur, Erwin." In *Encyclopedia of Genetics,* edited by Sydney Brenner and Jeffrey H. Miller, 199–203. London: Academic Press, 2001.

Lowe, Dunstan. "The Symbolic Value of Grafting in Ancient Rome." *Transactions of the American Philological Association* 140, no. 2 (2010): 461–88.

Lurquin, Paul F. *The Green Phoenix: A History of Genetically Modified Plants.* New York: Columbia University Press, 2001.

Lysenko, T. D. *Agrobiology: Essays on Problems of Genetics, Plant Breeding and Seed Growing*. Moscow: Foreign Languages Publishing House, 1954.

MacBride, E. W. "Variety and Environment in Lizards." *Nature* 120, no. 3011 (1927): 71–74.

MacDowell, E. Carleton. "Charles Benedict Davenport, 1866–1944: A Study of Conflicting Influences." *Bios* 17, no. 1 (1946): 3–50.

Macfarlane, J. Muirhead. "A Comparison of the Minute Structure of Plant Hybrids with That of Their Parents, and Its Bearing on Biological Problems." *Transactions of the Royal Society of Edinburgh* 37, no. 14 (1895): 203–86.

Macfarlane, J. Muirhead. "Observations on Some Hybrids between *Drosera filiformis* and *D. intermedia*." *Journal of the Royal Horticultural Society*, 24 (1900): 241–49.

Magner, Lois N. Review of *Beyond the Gene: Cytoplasmic Inheritance and the Struggle for Authority in Genetics*, by Jan Sapp. *American Historical Review* 95, no. 2 (1990): 451–52.

Maienschein, Jane. "On Cloning: Advocating History of Biology in the Public Interest." *Journal of the History of Biology* 34, no. 3 (2001): 423–32.

Mather, Kenneth. "Perspective and Prospect." *Philosophical Transactions of the Royal Society of London. Series B, Biological Sciences* 292, no. 1062 (1981): 601–9.

McIntosh, Charles. *The Book of the Garden*. Vol. 2. Edinburgh: William Blackwood and Sons, 1853.

McLaren, Anne. "International Rapprochement, 50 Years Ago." *Transplantation* 76, no. 10 (2003): 1425.

McLaren, Anne. "Too Late for the Midwife Toad: Stress, Variability and Hsp90." *Trends in Genetics* 15, no. 5 (1999): 169–71.

McLaren, Anne, and Donald Michie. "Current Trends of Genetical Research in Hungary." *Nature* 174, no. 4426 (1954): 390–91.

Medawar, Peter B. "Is the Scientific Paper a Fraud?" *Listener* 70, no. 12 (1963): 377–78.

Medawar, Peter B. *Memoir of a Thinking Radish: An Autobiography*. Oxford: Oxford University Press, 1986.

Mei, Jianjun. "Some Reflections on Joseph Needham's Intellectual Heritage." *Technology and Culture* 60, no. 2 (2019): 594–603.

Meloni, Maurizio. *Political Biology: Science and Social Values in Human Heredity from Eugenics to Epigenetics*. Basingstoke: Palgrave Macmillan, 2016.

Michel, W. "Uber die experimentelle Fusion pflanzlicher Protoplasten." *Archiv für experimentelle Zellforschung besonders Gewebezüchtung* 20 (1937): 230–52.

Michie, D., and A. McLaren. "The Importance of Being Cross-Bred." *New Biology* 19 (1955): 48–69.

Michurin, Ivan Vladimirovich. *Selected Works*. Moscow: Foreign Languages Publishing House, 1949.

Michurin, Ivan Vladimirovich. *Some Problems of Method*. Moscow: Foreign Languages Publishing House, 1952.

Mougey, Thomas. "Needham at the Crossroads: History, Politics and International Science in Wartime China (1942–1946)." *British Journal for the History of Science* 50, no. 1 (2017): 83–109.

Mudge, Ken, Jules Janick, Steven Scofield, and Eliezer E. Goldschmidt. "A History of Grafting." *Horticultural Reviews* 35 (2009): 437–93.

Muller, Hermann Joseph. "Artificial Transmutation of the Gene." *Science* 66, no. 1699 (1927): 84–87.

Müller-Wille, Staffan, and Giuditta Parolini. "Punnett Squares and Hybrid Crosses: How Mendelians Learned Their Trade by the Book." *BJHS Themes* 5 (2020): 149–65.

Müller-Wille, Staffan, and Hans-Jörg Rheinberger. *A Cultural History of Heredity*. Chicago: University of Chicago Press, 2012.

Munns, David P. D. "The Phytotronist and the Phenotype: Plant Physiology, Big Science, and a Cold War Biology of the Whole Plant." *Studies in History and Philosophy of Science. Part C, Studies in History and Philosophy of Biological and Biomedical Sciences* 50 (2015): 29–40.

Nagata, Toshiyuki, and Itaru Takebe. "Cell Wall Regeneration and Cell Division in Isolated Tobacco Mesophyll Protoplasts." *Planta* 92, no. 4 (1970): 301–8.

Needham, Joseph, Gwei-Djen Lu, and H. T. Huang. *Science and Civilisation in China.* Vol. 6, *Biology and Biological Technology: Part 1, Botany.* Cambridge: Cambridge University Press, 1986.

Neilson Jones, W. "Chimaeras: A Summary and Some Special Aspects." *Botanical Review* 3, no. 11 (1937): 545–62.

Neilson Jones, W. *The Growing Plant.* London: Faber and Faber, 1948.

Neilson Jones, W. *Plant Chimaeras and Graft Hybrids.* London: Methuen, 1934.

Neilson Jones, W., and M. C. Rayner. *Textbook of Plant Biology.* London: Methuen, 1920.

Nickell, L. G., and J. G. Torrey. "Crop Improvement through Plant Cell and Tissue Culture." *Science* 166, no. 3908 (1969): 1068–70.

Ohta, Yasuo. "Graft-Transformation, the Mechanism for Graft-Induced Genetic Changes in Higher Plants." *Euphytica* 55 (1991): 91–99.

Ohta, Yasuo. "Tsugiki de kawaru iden keishitsu" [Hereditary changes induced by grafting]. *Saiensu* [Scientific American (Japanese ed.)] 7, no. 7 (1977): 100–113.

Ohta, Yasuo. "A Variant Found in the Progeny from Grafting in *Capsicum annuum*." *National Institute of Genetics, Japan: Annual Report* 20 (1970): 34–35.

Ohta, Yasuo, and Phan Van Chuong, "Hereditary Changes in *Capsicum Annuum* L. I. Induced by Ordinary Grafting." *Euphytica* 24, no. 2 (1975) 355–68.

Okada, Yoshio. "Cell Fusion and Somatic Cell Genetics." *Japanese Journal of Human Genetics* 24, no. 3 (1979): 143–44.

Olby, Robert C. "Bateson, William." In *Complete Dictionary of Scientific Biography*, edited by Charles Coulston Gillispie, vol. 1, *Pierre Abailard–L. S. Berg*, 505–6. Detroit: Charles Scribner's Sons, 2008.

Olby, Robert C. "Mendel No Mendelian?" *History of Science* 17, no. 1 (1979): 53–72.

Olby, Robert C. *Origins of Mendelism.* 2nd ed. Chicago: University of Chicago Press, 1985.

Olmstead, Alan L., and Paul W. Rhode. *Creating Abundance: Biological Innovation and American Agricultural Development.* Cambridge: Cambridge University Press, 2008.

Onaga, Lisa. "Toyama Kametaro and Vernon Kellogg: Silkworm Inheritance Experiments in Japan, Siam, and the United States, 1900–1912." *Journal of the History of Biology* 43, no. 2 (2010): 215–64.

Paleček, Pavel. "Vítězslav Orel (1926–2015): Gregor Mendel's Biographer and the Rehabilitation of Genetics in the Communist Bloc." *History and Philosophy of the Life Sciences* 38, no. 3 (2016): 1–12.

Palladino, Paolo. *Plants, Patients and the Historian: (Re)membering in the Age of Genetic Engineering.* Manchester: Manchester University Press, 2002.

Palló, Gábor, and Miklós Müller. "Opportunism and Enforcement: Hungarian Reception of Michurinist Biology in the Cold War Period." In *The Lysenko Controversy as a Global Phenomenon*, edited by William deJong-Lambert and Nikolai Krementsov, Vol. 2, *Genetics and Agriculture in the Soviet Union and Beyond*, 3–36. Cham, Switzerland: Palgrave Macmillan, 2017.

Paul, Diane B. "A War on Two Fronts: J. B. S. Haldane and the Response to Lysenkoism in Britain." *Journal of the History of Biology* 16, no. 1 (1983): 1–37.

Paul, Diane B., and Barbara A. Kimmelman. "Mendel in America: Theory and Practice, 1900–1919." In *The American Development of Biology*, edited by Ronald Rainger, Keith R. Benson, and Jane Maienschein, 281–310. Philadelphia: University of Pennsylvania Press, 1988.

Pauly, Philip J. *Controlling Life: Jacques Loeb and the Engineering Ideal in Biology*. New York: Oxford University Press, 1987.

Pauly, Philip J. *Fruits and Plains: The Horticultural Transformation of America*. Cambridge, MA: Harvard University Press, 2007.

Pawley, Emily. *The Nature of the Future: Agriculture, Science, and Capitalism in the Antebellum North*. Chicago: University of Chicago Press, 2020.

Pease, Arthur Stanley. "Notes on Ancient Grafting." *Transactions and Proceedings of the American Philological Association* 64 (1933): 66–76.

Pental, Deepak, and Edward C. Cocking. "Some Theoretical and Practical Possibilities of Plant Genetic Manipulation using Protoplasts." *Hereditas* 103, suppl. 3 (1985): 83–92.

Plowe, Janet Q. "Membranes in the Plant Cell. I. Morphological Membranes at Protoplasmic Surfaces." *Protoplasma* 12 (1931): 196–220.

Poiteau, P. A. "Remarks on the *Cytisus Adami*, or Purple Laburnum." *Gardener's Magazine* 17 (1841): 58–61.

Popenoe, Paul Bowman. "Plant Chimeras: Recent Spectacular Productions of Experimental Horticulture." *Journal of Heredity* 5, no. 12 (1914): 520–32.

Porter, Helen K. "Vernon Herbert Blackman, 1872–1967." *Biographical Memoirs of Fellows of the Royal Society* 14 (1968): 37–60.

Power, J. B., E. M. Frearson, C. Hayward, and E. C. Cocking. "Some Consequences of the Fusion and Selective Culture of *Petunia* and *Parthenocissus* Protoplasts." *Plant Science Letters* 5, no. 3 (1975): 197–207.

Pringle, Peter. *The Murder of Nikolai Vavilov: The Story of Stalin's Persecution of One of the Twentieth Century's Greatest Scientists*. London: JR Books, 2009.

"Problems of Genetics." *Nature* 92, no. 2305 (1914): 497–98.

R.P.G. "Graft-Hybrids." *New Phytologist* 10, no. 5–6 (1911): 212–15.

Radick, Gregory. *Disputed Inheritance: The Battle over Mendel and the Future of Biology*. Chicago: University of Chicago Press, 2023.

Radick, Gregory. "Other Histories, Other Biologies." In *Philosophy, Biology and Life*, edited by Anthony O'Hear, 21–47. Cambridge: Cambridge University Press, 2005.

Ragionieri, Ritratto di Attilio. "Origin of the Florentine Bizzarria." *Journal of Heredity* 18, no. 12 (1927): 527–28.

Rasmussen, Nicolas. *Gene Jockeys: Life Science and the Rise of Biotech Enterprise*. Baltimore, MD: Johns Hopkins University Press, 2014.

Richmond, Marsha L. "Women in the Early History of Genetics: William Bateson and the Newnham College Mendelians, 1900–1910." *Isis* 92, no. 1 (2001): 55–90.

Risso, A., and A. Poiteau. *Histoire naturelle des orangers*. Paris: Mme Hérissant Le Doux, 1818.

Roberts, J. A. Fraser. "Reginald Ruggles Gates, 1882–1962." *Biographical Memoirs of Fellows of the Royal Society* 10 (1964): 83–106.

Roberts, R. H. "Theoretical Aspects of Graftage." *Botanical Review* 15, no. 7 (1949): 423–63.

Robertson, Thomas. *The Malthusian Moment: Global Population Growth and the Birth of American Environmentalism*. New Brunswick, NJ: Rutgers University Press, 2012.

Roll-Hansen, Nils. "Lamarckism and Lysenkoism Revisited." In *Transformations of Lamarckism: From Subtle Fluids to Molecular Biology*, edited by Snait B. Gissis and Eva Jablonka, 77–88. Cambridge, MA: MIT Press, 2011.

Roll-Hansen, Nils. *The Lysenko Effect: The Politics of Science*. Amherst, MA: Humanity Books, 2006.

Roll-Hansen, Nils. "A New Perspective on Lysenko?" *Annals of Science* 42, no. 3 (1985): 261–78.

Rossianov, Kirill O. "Editing Nature: Joseph Stalin and the 'New' Soviet Biology." *Isis* 84, no. 4 (1993): 728–45.

Rushton, Alan R. "William Bateson and the Chromosome Theory of Heredity: A Reappraisal." *British Journal for the History of Science* 47, no. 1 (2014): 147–71.

Sachs, L. "Vegetative Hybridization." *Nature* 164, no. 4180 (1949): 1009–10.

Saito, Hirofumi. "Why Did Japanese Geneticists Take a Scientific Interest in Lysenko's Theories?" In *The Lysenko Controversy as a Global Phenomenon*, edited by William deJong-Lambert and Nikolai Krementsov, vol. 2, *Genetics and Agriculture in the Soviet Union and Beyond*, 137–57. Cham, Switzerland: Palgrave Macmillan, 2017.

Salisbury, E. J. Review of *Plant Chimæras and Graft Hybrids*, by W. Neilson Jones. *Science Progress* 30, no. 117 (1935): 173.

Sapp, Jan. *Beyond the Gene: Cytoplasmic Inheritance and the Struggle for Authority in Genetics*. New York: Oxford University Press, 1987.

Sapp, Jan. "The Struggle for Authority in the Field of Heredity, 1900–1932: New Perspectives on the Rise of Genetics." *Journal of the History of Biology* 16, no. 3 (1983): 311–42.

Savoia, Paolo. "Nature or Artifice? Grafting in Early Modern Surgery and Agronomy." *Journal of the History of Medicine and Allied Sciences* 72, no. 1 (2017): 67–86.

Schneider, Daniel. *Hybrid Nature: Sewage Treatment and the Contradictions of the Industrial Ecosystem.* Cambridge, MA: MIT Press, 2011.

Schneider, Laurence. *Biology and Revolution in Twentieth-Century China.* Lanham, MD: Rowman and Littlefield, 2003.

Schneider, Laurence. "Michurinist Biology in the People's Republic of China, 1948–1956." *Journal of the History of Biology* 45, no. 3 (2012): 525–56.

Schoijet, Mauricio. "*Limits to Growth* and the Rise of Catastrophism." *Environmental History* 4, no. 4 (1999): 515–30.

Schurman, Rachel A. "Biotechnology in the New Millennium: Technological Change, Institutional Change, and Political Struggle." In *Engineering Trouble: Biotechnology and Its Discontents*, edited by Rachel A. Schurman and Dennis Doyle Takahashi Kelso, 1–23. Berkeley: University of California Press, 2003.

Schwartz, Joel S. "George John Romanes's Defense of Darwinism: The Correspondence of Charles Darwin and His Chief Disciple." *Journal of the History of Biology* 28, no. 2 (1995): 281–316.

Scofield, Carl S. "Description Forms and Score Cards as Helps to Breeders." *Journal of Heredity*, o.s., 1, no. 1 (1905): 24–29.

Scowcroft, W. R. "Genetic Manipulation in Crops: A Symposium Review." in *Genetic Manipulation in Crops: Proceedings of the International Symposium . . . Beijing, October 1984*, edited by the International Rice Research Institute, 13–17. London: Cassell Tycooly, 1988.

Shepard, James F., Dennis Bidney, Tina Barsby, and Roger Kemble. "Genetic Transfer in Plants through Interspecific Protoplast Fusion." *Science* 219, no. 4585 (1983): 683–88.

Shull, George H. "Graft-Hybrids." *Botanical Gazette* 41, no. 5 (1906): 358–59.

Shull, G. H. "The 'Graft Hybrids' of Bronvaux." *Botanical Gazette* 60, no. 4 (1915): 323–24.

Simmonds, Norman W., and J. Smartt, eds. *Principles of Crop Improvement*. 2nd ed. Oxford: Blackwell Science, 1999.

Simpson, Q. I., and J. P. Simpson. "Genetic Laws Applied." *Journal of Heredity*, o.s., 5, no. 1 (1909): 250–55.

Sirks, M. J. "The Royal Horticultural Society and the Science of Genetics." *Journal of the Royal Horticultural Society* 80 (1955): 214–19.

Skene, Macgregor. "Plant Chimæras." *Science Progress in the Twentieth Century (1906–1916)* 9, no. 33 (1914): 127–34.

Small, J. Review of *Plant Chimaeras and Graft Hybrids*, by W. Neilson Jones. *Irish Naturalists' Journal* 5, no. 4 (1934): 94.

Smith, Gilbert M. "Douglas Houghton Campbell, 1859–1953." *Biographical Memoirs of the National Academy of Sciences* (1956): 45–63.

Smith, Hugh B. "Chromosome Counts in the Varieties of *Solanum tuberosum* and Allied Wild Species." *Genetics* 12, no. 1 (1927): 84–92.

Solberg, Winton U. *Reforming Medical Education: The University of Illinois College of Medicine, 1880–1920*. Urbana: University of Illinois Press, 2009.

Spillman, William J. "Mendel's Law in Relation to Animal Breeding." *Journal of Heredity*, o.s., 1, no. 2 (1905): 171–77.

Stamhuis, Ida H. "Why the Rediscoverer Ended Up on the Sidelines: Hugo De Vries's Theory of Inheritance and the Mendelian Laws." *Science and Education* 24, no. 1–2 (2015): 29–49.

Stegemann, Sandra, and Ralph Bock. "Exchange of Genetic Material between Cells in Plant Tissue Grafts." *Science* 324, no. 5927 (2009): 649–51.

Stegemann, Sandra, Mandy Keuthe, Stephan Greiner, and Ralph Bock. "Horizontal Transfer of Chloroplast Genomes between Plant Species." *Proceedings of the National Academy of Sciences* 109, no. 7 (2012): 2434–38.

Stráner, Katalin. "The Natural Sciences and Their Public at the Meetings of the Hungarian Association for the Advancement of Science in Budapest and Beyond, 1841–1896." In *Urban Histories of Science: Making Knowledge in the City, 1820–1940*, edited by Oliver Hochadel and Agustí Nieto-Galan, 59–79. New York: Routledge, 2018.

Subramanian, Samanth. *A Dominant Character: The Radical Science and Restless Politics of J. B. S. Haldane*. London: Atlantic Books, 2020.

Sumner, James. *Brewing Science, Technology and Print, 1700–1880*. Pittsburgh: University of Pittsburgh Press, 2016.

Swingle, Charles F. "Graft Hybrids in Plants." *Journal of Heredity* 18, no. 2 (1927): 73–94.

Takebe, I., Gudrun Labib, and G. Melchers. "Regeneration of Whole Plants from Isolated Mesophyll Protoplasts of Tobacco." *Naturwissenschaften* 58 (1971): 318–20.

Tanaka, Tyôzaburô. "Bizzarria—A Clear Case of Periclinal Chimera." *Journal of Genetics* 18, no. 1 (1927): 77–85.

Taschwer, Klaus. *Der Fall Paul Kammerer: Das abenteuerliche Leben des umstrittensten Biologen seiner Zeit*. Munich: Carl Hanser, 2016.

Thomas, H. Hamshaw. "Frederick Ernest Weiss, 1865–1953." *Biographical Memoirs of Fellows of the Royal Society* 8, no. 22 (1953): 601–8.

Thurtle, Phillip. *The Emergence of Genetic Rationality: Space, Time, and Information in American Biological Science, 1870–1920*. Seattle: University of Washington Press, 2007.

Tilney-Bassett, Richard A. E. *Plant Chimeras*. London: Edward Arnold, 1986.

Tompsett, Ben. "A Fruit Grower Visits the U.S.S.R." *Journal of the Royal Horticultural Society* 94 (1969): 354–62.

Tsu, Teh-Ming, and Yu-Seng Chao. "A Study on the Vegetative Hybridization of Some Solanaceous Plants." *Scientia Sinica* 6, no. 5 (1957): 889–903.

van Dijk, Peter J., Franz J. Weissing, and T. H. Noel Ellis. "How Mendel's Interest in Inheritance Grew out of Plant Improvement." *Genetics* 210, no. 2 (2018): 347–55.

von Schwerin, Alexander. "Seeing, Breeding and the Organisation of Variation: Erwin Baur and the Culture of Mutations in the 1920s." In *Conference: Heredity in the Century of the Gene (A Cultural History of Heredity IV)*, preprint 343, 259–78. Berlin: Max Planck Institute for the History of Science, 2008.

Wardy, Robert. "The Mysterious Aristotelian Olive." *Science in Context* 18, no. 1 (2005): 69–91.

Weiner, Douglas R. "The Roots of 'Michurinism': Transformist Biology and Acclimatization as Currents in the Russian Life Sciences." *Annals of Science* 42, no. 3 (1985): 243–60.

Weiss, F. E. "Graft Hybrids." *Annual Report and Transactions, Manchester Microscopical Society* (1916): 31–41.

Weiss, F. E. "Graft Hybrids and Chimaeras: I." *Journal of the Royal Horticultural Society* 64, no. 7 (1940): 212–17.

Weiss, F. E. "Graft Hybrids and Chimaeras: II." *Journal of the Royal Horticultural Society* 65, no. 1 (1940): 237–43.

Weiss, F. E. "Life." *Annual Report and Transactions, Manchester Microscopical Society* (1898): 64–76.

Weiss, F. E. "Microscopy in Manchester." *Annual Report and Transactions, Manchester Microscopical Society* (1930): 36–54.

Weiss, F. E. "On the Leaf-Tissues of the Graft Hybrids *Cratægo-Mespilus asniersii* and *Cratægo-Mespilus dardari*." *Manchester Memoirs* 69, no. 9 (1924–25): 73–78.

Weiss, F. E. "The Problem of Graft Hybrids and Chimaeras." *Biological Reviews* 5, no. 3 (1930): 231–71.

Weiss, F. E. "Researches on Heredity in Plants." *Memoirs and Proceedings of the Manchester Literary and Philosophical Society* 56, pt. 1. (1912): 1–12.

Weiss, F. E. Review of *Plant Chimaeras and Graft Hybrids*, by W. Neilson Jones. *New Phytologist* 33, no. 5 (1934): 390–91.

Weiss, F. E. "Some Recent Advances in Our Knowledge of Inheritance in Plants." *Manchester Memoirs* 71, no. 8 (1926–27): 75–86.

Werskey, Gary. *The Visible College: A Collective Biography of British Scientists and Socialists of the 1930s.* London: Free Association Books, 1988.

Werskey, Gary. "The Visible College Revisited: Second Opinions on the Red Scientists of the 1930s." *Minerva* 45, no. 3 (2007): 305–19.

Wharton, Clifton R., Jr. "Food, the Hidden Crisis." *Science* 208, no. 4451 (1980): 1415.

Whittaker, J. R. "Siphon Regeneration in *Ciona*." *Nature* 255, no. 5505 (1975): 224–25.

Williams, Nicola. "Irene Manton, Erwin Schrödinger and the Puzzle of Chromosome Structure." *Journal of the History of Biology* 49, no. 3 (2016): 425–59.

Wilmot, Sarah. "J. B. S. Haldane: The John Innes Years." *Journal of Genetics* 96, no. 5 (2017): 815–26.

Wilson, Duncan. *Tissue Culture in Science and Society: The Public Life of a Biological Technique in Twentieth Century Britain.* Basingstoke: Palgrave Macmillan, 2011.

Winther, Rasmus G. "August Weismann on Germ-Plasm Variation." *Journal of the History of Biology* 34, no. 3 (2001): 517–55.

Witkowski, Jan A. "Charles Benedict Davenport, 1866–1944." In *Davenport's Dream: 21st Century Reflections on Heredity and Eugenics*, edited by Jan A. Witkowski and John R. Inglis, 35–58. New York: Cold Spring Harbor Laboratory Press, 2008.

Witkowski, Jan A. Review of *Beyond the Gene: Cytoplasmic Inheritance and the Struggle for Authority in Genetics*, by Jan Sapp. *Medical History* 34, no. 2 (1990): 233–34.

Wolfe, Audra J. "The Cold War Context of the Golden Jubilee, or, Why We Think of Mendel as the Father of Genetics." *Journal of the History of Biology* 45, no. 3 (2012): 389–414.

Wolfe, Audra J. "What Does It Mean to Go Public? The American Response to Lysenkoism, Reconsidered." *Historical Studies in the Natural Sciences* 40, no. 1 (2010): 48–78.

Wood, Roger J., and Vítěslav Orel. *Genetic Prehistory in Selective Breeding: A Prelude to Mendel*. Oxford: Oxford University Press, 2001.

Yang, Lei, Frank Machin, Shuangfeng Wang, Eleftheria Saplaoura, and Friedrich Kragler. "Heritable Transgene-Free Genome Editing in Plants by Grafting of Wild-Type Shoots to Transgenic Donor Rootstocks." *Nature Biotechnology* (2023): 958–67.

Zallen, Doris T. "Redrawing the Boundaries of Molecular Biology: The Case of Photosynthesis." *Journal of the History of Biology* 26, no. 1 (1993): 65–87.

Zirkle, Conway. "The Role of Liberty Hyde Bailey and Hugo de Vries in the Rediscovery of Mendelism." *Journal of the History of Biology* 1, no. 2 (1968): 205–18.

INDEX

Note: Page references in *italics* refer to figures.